Strip a Man of His Clothes and Conventions and What Do You Find?

"If one took a group of twenty suburban families and placed them in a primitive sub-tropical environment where the males had to go off hunting for food, the sexual structure of this new tribe would require little or no modification. In fact, what has happened in every large town or city is that the individuals it contains have specialized their hunting (working) techniques, but have retained their socio-sexual system in more or less its original form. . . . Only in the field of general breeding information are we now coming face to face with the first major assault on our age-old sexual system by the forces of modern civilization. Thanks to medical science, surgery and hygiene, we have reached an incredible peak of breeding success. We have practised death control and now we must balance it with birth control. It looks very much as though, during the next century or so, we are going to change our sexual ways at last. But if we do, it will not be because they failed, but because they succeeded too well."

—from *The Naked Ape*

"This is not only a thoughtful and stimulating book, but also an extremely interesting one."

—*The Times* (of London) *Literary Supplement*

THE NAKED APE

DESMOND MORRIS

*A Zoologist's Study of the
Human Animal*

Delta
Trade Paperbacks

A Delta Book
Published by
Dell Publishing
a division of
Random House, Inc.
1540 Broadway
New York, New York 10036

ISBN: 0-385-33430-3

Reprinted by arrangement with McGraw-Hill

Manufactured in the United States of America

Published simultaneously in Canada

April 1999

10 9 8 7 6

BVG

Preface to the new edition of *The Naked Ape,* 1983

This book was written in November 1966 in four explosive, exhausting weeks. I can still recall the surprise I felt when I discovered that the palms of my hands were sweating as I typed. There was a manic intensity about those four weeks which at the time I could not explain. Looking back, it is easier to understand.

The writing of *The Naked Ape* was a short, sharp climax to a long gestation period. As a child I had always been surrounded by animals of many kinds—foxes, crows, rabbits, lizards, parrots, snakes, toads, shrews, newts, voles, hedgehogs—the list was endless. I kept them, watched them, studied them and bred them. I enjoyed their company more than that of other members of my own species.

Later, as a young professional zoologist, I started a major investigation into the behaviour of fish, following that a few years later by another long study of bird behaviour. Moving to the Zoological Society of London, I became Curator of Mammals and concentrated my researches on a wide variety of mammalian species, culminating with detailed studies of monkeys and apes, especially chimpanzees. It was as if I was unconsciously working my way up the family tree towards my own kind, approaching the human species, as it were, from below.

The childhood shyness that had driven me so much into the company of other animals was gone. I was now at home with humans, but having approached them in this roundabout way meant that I saw them in a rather unusual light. I viewed my fellow-man not as a fallen angel, but as a risen ape—a naked ape of remarkable resilience, energy and imagination, but an animal for all that. Just another species for me to examine.

I knew that many people resented being called animals, as though this was in some way disgusting—an insult to human dignity. Since I had always loved animals I found this rather depressing. It meant that such people had a low opinion of the other members of the animal kingdom. For me, it was just the reverse. Before I wrote *The Naked Ape,* I had preferred to study other, more beautiful and more fascinating species. Now, at last, I was prepared to *elevate* the human species to the level of one of my beloved animal forms.

Of course, I guessed that I might shock some of the more starry-eyed escapists—people who were still gullible enough to believe the old fairy-tales designed to keep superstitious medieval peasants in their place—and I also suspected that the deliberate frankness of some of my statements might prove distasteful to the more sheltered puritans. But I was in no mood to compromise or to soften my message. I wanted to tell the truth as I saw it, bluntly and straightforwardly, with all the usual 'waffling', side-stepping and philosophical smoke-screening swept away. The ape was naked, not merely because he had lost his thick coat of fur during the course of evolution, but also because I intended to strip him bare on the pages of my book and show him as he really is—a remarkable, ingenious, brilliant animal. No less and no more.

CONTENTS

Acknowledgments

*

This book is intended for a general audience and authorities
have therefore not been quoted in the text. To do so would
have broken the flow of words and is a practice suitable only
for a more technical work. But many brilliantly original
papers and books have been referred to during the assembly
of this volume and it would be wrong to present it without
acknowledging their valuable assistance. At the end of the
book I have included a chapter-by-chapter appendix relating
the topics discussed to the major authorities concerned. This
appendix is then followed by a selected bibliography giving
the detailed references.

I would also like to express my debt and my gratitude to the
many colleagues and friends who have helped me, directly and
indirectly, in discussions, correspondence and many other
ways. They include, in particular, the following: Dr Anthony
Ambrose, Mr David Attenborough, Dr David Blest, Dr N. G.
Blurton-Jones, Dr John Bowlby, Dr Hilda Bruce, Dr
Richard Coss, Dr Richard Davenport, Dr Alisdair Fraser,
Professor J. H. Fremlin, Professor Robin Fox, Baroness Jane
van Lawick-Goodall, Dr Fae Hall, Professor Sir Alister Hardy,
Professor Harry Harlow, Mrs Mary Haynes, Dr Jan van
Hooff, Sir Julian Huxley, Miss Devra Kleiman, Dr Paul
Leyhausen, Dr Lewis Lipsitt, Mrs Caroline Loizos, Professor
Konrad Lorenz, Dr Malcolm Lyall-Watson, Dr Gilbert
Manley, Dr Isaac Marks, Mr Tom Maschler, Dr L. Harrison
Matthews, Mrs Ramona Morris, Dr John Napier, Mrs

7

Caroline Nicolson, Dr Kenneth Oakley, Dr Frances Reynolds, Dr Vernon Reynolds, The Hon. Miriam Rothschild, Mrs Claire Russell, Dr W. M. S. Russell, Dr George Schaller, Dr John Sparks, Dr Lionel Tiger, Pofessor Niko Tinbergen, Mr Ronald Webster, Dr Wolfgang Wickler, and Professor John Yudkin.

I hasten to add that the inclusion of a name in this list does not imply that the person concerned necessarily agrees with my views as expressed here in this book.

Introduction

*

There are one hundred and ninety-three living species of monkeys and apes. One hundred and ninety-two of them are covered with hair. The exception is a naked ape self-named *Homo sapiens*. This unusual and highly successful species spends a great deal of time examining his higher motives and an equal amount of time studiously ignoring his fundamental ones. He is proud that he has the biggest brain of all the primates, but attempts to conceal the fact that he also has the biggest penis, preferring to accord this honour falsely to the mighty gorilla. He is an intensely vocal, acutely exploratory, over-crowded ape, and it is high time we examined his basic behaviour.

I am a zoologist and the naked ape is an animal. He is therefore fair game for my pen and I refuse to avoid him any longer simply because some of his behaviour patterns are rather complex and impressive. My excuse is that, in becoming so erudite, *Homo sapiens* has remained a naked ape nevertheless; in acquiring lofty new motives, he has lost none of the earthy old ones. This is frequently a cause of some embarrassment to him, but his old impulses have been with him for millions of years, his new ones only a few thousand at the most—and there is no hope of quickly shrugging off the accumulated genetic legacy of his whole evolutionary past. He would be a far less worried and more fulfilled animal if only he would face up to this fact. Perhaps this is where the zoologist can help.

One of the strangest features of previous studies of naked-ape behaviour is that they have nearly always avoided the obvious. The earlier anthropologists rushed off to all kinds of unlikely corners of the world in order to unravel the basic truth about our nature, scattering to remote cultural back-waters so atypical and unsuccessful that they are nearly extinct. They then returned with startling facts about the bizarre mating customs, strange kinship systems, or weird ritual pro-cedures of these tribes, and used this material as though it were of central importance to the behaviour of our species as a whole. The work done by these investigators was, of course, extremely interesting and most valuable in showing us what can happen when a group of naked apes becomes side-tracked into a cultural blind alley. It revealed just how far from the normal our behaviour patterns can stray without a complete social collapse. What it did not tell us was anything about the typical behaviour of typical naked apes. This can only be done by examining the common behaviour patterns that are shared by all the ordinary, successful members of the major cultures —the mainstream specimens who together represent the vast majority. Biologically, this is the only sound approach. Against this, the old-style anthropologist would have argued that his technologically simple tribal groups are nearer the heart of the matter than the members of advanced civilizations. I submit that this is not so. The simple tribal groups that are living today are not primitive, they are stultified. Truly primitive tribes have not existed for thousands of years. The naked ape is essentially an exploratory species and any society that has failed to advance has in some sense failed, 'gone wrong'. Something has happened to it to hold it back, something that is working against the natural tendencies of the species to explore and investigate the world around it. The characteristics that the earlier anthropologists studied in these tribes may well be the very features that have interfered with the progress of the groups concerned. It is therefore dangerous to use this information as the basis for any general scheme of our behaviour as a species.

Psychiatrists and psycho-analysts, by contrast, have stayed nearer home and have concentrated on clinical studies of mainstream specimens. Much of their earlier material, although not suffering from the weakness of the anthropological information, also has an unfortunate bias. The individuals on which they have based their pronouncements are, despite their mainstream background, inevitably aberrant or failed specimens in some respect. If they were healthy, successful and therefore typical individuals, they would not have had to seek psychiatric aid and would not have contributed to the psychiatrists' store of information. Again, I do not wish to belittle the value of this research. It has given us an immensely important insight into the way in which our behaviour patterns can break down. I simply feel that in attempting to discuss the fundamental biological nature of our species as a whole, it is unwise to place too great an emphasis on the earlier anthropological and psychiatric findings.

(I should add that the situation in anthropology and psychiatry is changing rapidly. Many modern research workers in these fields are recognizing the limitations of the earlier investigations and are turning more and more to studies of typical, healthy individuals. As one investigator expressed it recently: 'We have put the cart before the horse. We have tackled the abnormals and we are only now beginning, a little late in the day, to concentrate on the normals.')

The approach I propose to use in this book draws its material from three main sources: (1) the information about our past as unearthed by palaeontologists and based on the fossil and other remains of our ancient ancestors; (2) the information available from the animal behaviour studies of the comparative ethologists, based on detailed observations of a wide range of animal species, especially our closest living relatives, the monkeys and apes; and (3) the information that can be assembled by simple, direct observation of the most basic and widely shared behaviour patterns of the successful mainstream specimens from the major contemporary cultures of the naked ape itself.

Because of the size of the task, it will be necessary to over-simplify in some manner. The way I shall do this is largely to ignore the detailed ramifications of technology and verbalization, and concentrate instead on those aspects of our lives that have obvious counterparts in other species: such activities as feeding, grooming, sleeping, fighting, mating and care of the young. When faced with these fundamental problems, how does the naked ape react? How do his reactions compare with those of other monkeys and apes? In which particular respect is he unique, and how do his oddities relate to his special evolutionary story?

In dealing with these problems I realize that I shall run the risk of offending a number of people. There are some who will prefer not to contemplate their animal selves. They may consider that I have degraded our species by discussing it in crude animal terms. I can only assure them that this is not my intention. There are others who will resent any zoological invasion of their specialist arena. But I believe that this approach can be of great value and that, whatever its short-comings, it will throw new (and in some ways unexpected) light on the complex nature of our extraordinary species.

Chapter One

*

ORIGINS

There is a label on a cage at a certain zoo that states simply, 'This animal is new to science'. Inside the cage there sits a small squirrel. It has black feet and it comes from Africa. No black-footed squirrel has ever been found in that continent before. Nothing is known about it. It has no name.

For the zoologist it presents an immediate challenge. What is it about its way of life that has made it unique? How does it differ from the three hundred and sixty-six other living species of squirrels already known and described? Somehow, at some point in the evolution of the squirrel family, the ancestors of this animal must have split off from the rest and established themselves as an independent breeding population. What was it in the environment that made possible their isolation as a new form of life? The new trend must have started out in a small way, with a group of squirrels in one area becoming slightly changed and better adapted to the particular conditions there. But at this stage they would still be able to inter-breed with their relatives nearby. The new form would be at a slight advantage in its special region, but it would be no more than a race of the basic species and could be swamped out, re-absorbed into the mainstream at any point. If, as time passed, the new squirrels became more and more perfectly tuned-in to their particular environment, the moment would eventually arrive when it would be advantageous for them to become isolated from possible contamination by their neighbours. At this stage their social and sexual behaviour would undergo special modifications, making inter-breeding with other kinds of squirrels unlikely and eventually impossible. At first, their anatomy may have changed and become better at coping with the special food of the district, but later their mating calls and displays would also differ, ensuring that they attract only mates of the new type. At last, a new species would have evolved, separate and discrete, a unique form of life, a three hundred and sixty-seventh kind of squirrel.

When we look at our unidentified squirrel in its zoo cage, we can only guess about these things. All we can be certain about is that the markings of its fur—its black feet—indicate that it is a new form. But these are only the symptoms, the rash that gives a doctor a clue about his patient's disease. To really understand this new species, we must use these clues only as a starting point, telling us there is something worth pursuing. We might try to guess at the animal's history, but that would be presumptuous and dangerous. Instead we will start humbly by giving it a simple and obvious label: we will call it the African black-footed squirrel. Now we must observe and record every aspect of its behaviour and structure and see how it differs from, or is similar to, other squirrels. Then, little by little, we can piece together its story.

The great advantage we have when studying such animals is that we ourselves are not black-footed squirrels—a fact which forces us into an attitude of humility that is becoming to proper scientific investigation. How different things are, how depressingly different, when we attempt to study the human animal. Even for the zoologist, who is used to calling an animal an animal, it is difficult to avoid the arrogance of subjective involvement. We can try to overcome this to some extent by deliberately and rather coyly approaching the human being as if he were another species, a strange form of life on the dissecting table, awaiting analysis. How can we begin?

As with the new squirrel, we can start by comparing him with other species that appear to be most closely related. From his teeth, his hands, his eyes and various other anatomical features, he is obviously a primate of some sort, but of a very odd kind. Just how odd becomes clear when we lay out in a long row the skins of the one hundred and ninety-two living species of monkeys and apes, and then try to insert a human pelt at a suitable point somewhere in this long series. Wherever we put it, it looks out of place. Eventually we are driven to position it right at one end of the row of skins, next to the hides of the tailless great apes such as the chimpanzee and the gorilla. Even here it is obtrusively different. The legs are too long, the arms are too short and the feet are rather strange. Clearly this species of primate has developed a special kind of locomotion which has modified its basic form. But there is another characteristic that cries out for attention: the skin is virtually naked. Except for conspicuous tufts of hair on the head, in the armpits and around the genitals, the skin surface is completely exposed. When compared with the other primate species, the contrast is dramatic. True, some species of monkeys and apes have small naked patches of skin on their rumps, their faces, or their chests, but nowhere amongst the other one hundred and ninety-two species is there anything even approaching the human condition. At this point and without further investigation, it is justifiable to name this new species the 'naked ape'. It is a simple, descriptive name based on a simple observation, and it makes no special assumptions. Perhaps it will help us to keep a sense of proportion and maintain our objectivity.

Staring at this strange specimen and puzzling over the significance of its unique features, the zoologist now has to start making comparisons. Where else is nudity at a premium? The other primates are no help, so it means looking farther afield. A rapid survey of the whole range of the living mammals soon proves that they are remarkably attached to their protective, furry covering, and that very few of the 4,237 species in existence have seen fit to abandon it. Unlike their reptilian ancestors, mammals have acquired the great physiological advantage of being able to maintain a constant, high body temperature. This keeps the delicate machinery of the body processes tuned in for top performance. It is not a property to be endangered or discarded lightly. The temperature-controlling devices are of vital importance and the possession of a thick, hairy, insulating coat obviously plays a major role in preventing heat loss. In intense sunlight it will also prevent over-heating and damage to the skin from direct exposure to the sun's rays. If the hair has to go, then clearly there must be a very powerful reason for abolishing it. With few exceptions this drastic step has been taken only when mammals have launched themselves into an entirely new medium. The flying mammals, the bats, have been forced to denude their wings, but they have retained their furriness elsewhere and can hardly be counted as naked species. The burrowing mammals have in a few cases—the naked mole rat, the aardvark and the armadillo, for example—reduced their hairy covering. The aquatic mammals such as the whales, dolphins, porpoises, dugongs, manatees and hippopotamuses have also gone naked as part of a general streamlining. But for all the more typical surface-dwelling mammals, whether scampering about on the ground or clambering around in the vegetation, a densely hairy hide is the basic rule. Apart from those abnormally heavy giants, the rhinos and the elephants (which have heating and cooling problems peculiar to themselves), the naked ape stands alone, marked off by his nudity from all the thousands of hairy, shaggy or furry land-dwelling mammalian species.

At this point the zoologist is forced to the conclusion that either he is dealing with a burrowing or an aquatic mammal, or there is something very odd, indeed unique, about the evolutionary history of the naked ape. Before setting out on a field trip to observe the animal in its present-day form, the first thing to do, then, is to dig back into its past and examine as closely as possible its immediate ancestors. Perhaps by examining the fossils and other remains and by taking a look at the closest living relatives, we shall be able to gain some sort of picture of what happened as this new type of primate emerged and diverged from the family stock.

It would take too long to present here all the tiny fragments of evidence that have been painstakingly collected over the past century. Instead, we will assume that this task has been done and simply summarize the conclusions that can be drawn from it, combining the information available from the work of the fossil-hungry palaeontologists with the facts gathered by the patient ape-watching ethologists.

The primate group, to which our naked ape belongs, arose originally from primitive insectivore stock. These early mammals were small, insignificant creatures, scuttling nervously around in the safety of the forests, while the reptile overlords were dominating the animal scene. Between eighty and fifty million years ago, following the collapse of the great age of reptiles, these little insect-eaters began to venture out into new territories. There they spread and grew into many strange shapes. Some became plant-eaters and burrowed under the ground for safety, or grew long, stilt-like legs with which to flee from their enemies. Others became long-clawed, sharp-toothed killers. Although the major reptiles had abdicated and left the scene, the open country was once again a battlefield.

Meanwhile, in the undergrowth, small feet were still clinging to the security of the forest vegetation. Progress was being
made here, too. The early insect-eaters began to broaden their
diet and conquer the digestive problems of devouring fruits,
nuts, berries, buds and leaves. As they evolved into the lowliest
forms of primates, their vision improved, the eyes coming
forward to the front of the face and the hands developing as
food-graspers. With three-dimensional vision, manipulating
limbs and slowly enlarging brains, they came more and more
to dominate their arboreal world.

Somewhere between twenty-five and thirty-five million
years ago, these pre-monkeys had already started to evolve
into monkeys proper. They were beginning to develop long,
balancing tails and were increasing considerably in body size.
Some were on their way to becoming leaf-eating specialists,
but most were keeping to a broad, mixed diet. As time passed,
some of these monkey-like creatures became bigger and
heavier. Instead of scampering and leaping they switched to
brachiating — swinging hand over hand along the underside of
the branches. Their tails became obsolete. Their size, although
making them more cumbersome in the trees, made them less
wary of ground-level sorties.

Even so, at this stage — the ape phase — there was much to be
said for keeping to the lush comfort and easy pickings of their
forest of Eden. Only if the environment gave them a rude
shove into the great open spaces would they be likely to move.
Unlike the early mammalian explorers, they had become
specialized in forest existence. Millions of years of development had gone into perfecting this forest aristocracy, and if
they left now they would have to compete with the (by this
time) highly advanced ground-living herbivores and killers.
And so there they stayed, munching their fruit and quietly
minding their own business.

It should be stressed that this ape trend was for some reason taking place only in the Old World. Monkeys had evolved separately as advanced tree-dwellers in both the Old and the New World, but the American branch of the primates never made the ape grade. In the Old World, on the other hand, ancestral apes were spreading over a wide forest area from western Africa, at one extreme, to south-eastern Asia at the other. Today the remnants of this development can be seen in the African chimpanzees and gorillas and the Asian gibbons and orang-utans. Between these two extremities the world is now devoid of hairy apes. The lush forests have gone.

What happened to the early apes? We know that the climate began to work against them and that, by a point somewhere around fifteen million years ago, their forest strongholds had become seriously reduced in size. The ancestral apes were forced to do one of two things: either they had to cling on to what was left of their old forest homes, or, in an almost biblical sense, they had to face expulsion from the Garden. The ancestors of the chimpanzees, gorillas, gibbons and orangs stayed put, and their numbers have been slowly dwindling ever since. The ancestors of the only other surviving ape – the naked ape – struck out, left the forests, and threw themselves into competition with the already efficiently adapted ground-dwellers. It was a risky business, but in terms of evolutionary success it paid dividends.

The naked ape's success story from this point on is well known, but a brief summary will help, because it is vital to keep in mind the events which followed if we are to gain an objective understanding of the present-day behaviour of the species.

Faced with a new environment, our ancestors encountered a bleak prospect. They had to become either better killers than the old-time carnivores, or better grazers than the old-time herbivores. We know today that, in a sense, success has been won on both scores; but agriculture is only a few thousand years old, and we are dealing in millions of years. Specialized exploitation of the plant life of the open country was beyond the capacity of our early ancestors and had to await the development of advanced techniques of modern times. The digestive system necessary for a direct conquest of the grass-land food supply was lacking. The fruit and nut diet of the forest could be adapted to a root and bulb diet at ground level, but the limitations were severe. Instead of lazily reaching out to the end of the branch for a luscious ripe fruit, the vegetable-seeking ground ape would be forced to scratch and scrape painstakingly in the hard earth for his precious food.

His old forest diet, however, was not all fruit and nut. Animal proteins were undoubtedly of great importance to him. He came originally, after all, from basic insectivore stock, and his ancient arboreal home had always been rich in insect life. Juicy bugs, eggs, young helpless nestlings, tree-frogs and small reptiles were all grist to his mill. What is more, they posed no great problems for his rather generalized digestive system. Down on the ground this source of food supply was by no means absent and there was nothing to stop him increasing this part of his diet. At first, he was no match for the professional killer of the carnivore world. Even a small mongoose, not to mention a big cat, could beat him to the kill. But young animals of all kinds, helpless ones or sick ones, were there for the taking, and the first step on the road to major meat-eating was an easy one. The really big prizes, however, were poised on long, stilt-like legs, ready to flee at a moment's notice at quite impossible speeds. The protein-laden ungulates were beyond his grasp.

This brings us to the last million or so years of the naked ape's ancestral history, and to a series of shattering and increasingly dramatic developments. Several things happened together, and it is important to realize this. All too often, when the story is told, the separate parts of it are spread out as if one major advance led to another, but this is misleading. The ancestral ground-apes already had large and high-quality brains. They had good eyes and efficient grasping hands. They inevitably, as primates, had some degree of social organization. With strong pressure on them to increase their prey-killing prowess, vital changes began to take place. They became more upright—fast, better runners. Their hands became freed from locomotion duties—strong, efficient weapon-holders. Their brains became more complex—brighter, quicker decision-makers. These things did not follow one another in a major, set sequence; they blossomed together, minute advances being made first in one quality and then in another, each urging the other on. A hunting ape, a killer ape, was in the making.

It could be argued that evolution might have favoured the less drastic step of developing a more typical cat- or dog-like killer, a kind of cat-ape or dog-ape, by the simple process of enlarging the teeth and nails into savage fang-like and claw-like weapons. But this would have put the ancestral ground-ape into direct competition with the already highly specialized cat and dog killers. It would have meant competing with them on their own terms, and the outcome would no doubt have been disastrous for the primates in question. (For all we know, this may actually have been tried and failed so badly that the evidence has not been found.) Instead, an entirely new approach was made, using artificial weapons instead of natural ones, and it worked.

From tool-using to tool-making was the next step, and alongside this development went improved hunting techniques, not only in terms of weapons, but also in terms of social co-operation. The hunting apes were pack-hunters, and as their techniques of killing were improved, so were their methods of social organization. Wolves in a pack deploy themselves, but the hunting ape already had a much better brain than a wolf and could turn it to such problems as group communication and co-operation. Increasingly complex manœuvres could be developed. The growth of the brain surged on.

Essentially this was a hunting-group of males. The females were too busy rearing the young to be able to play a major role in chasing and catching prey. As the complexity of the hunt increased and the forays became more prolonged, it became essential for the hunting ape to abandon the meandering, nomadic ways of its ancestors. A home base was necessary, a place to come back to with the spoils, where the females and young would be waiting and could share the food. This step, as we shall see in later chapters, has had profound effects on many aspects of the behaviour of even the most sophisticated naked apes of today.

So the hunting ape became a territorial ape. His whole sexual, parental and social pattern began to be affected. His old wandering, fruit-plucking way of life was fading rapidly. He had now really left his forest of Eden. He was an ape with responsibilities. He began to worry about the prehistoric equivalent of washing machines and refrigerators. He began to develop the home comforts—fire, food storage, artificial shelters. But this is where we must stop for the moment, for we are moving out of the realms of biology and into the realms of culture. The biological basis of these advanced steps lies in the development of a brain large and complex enough to enable the hunting ape to take them, but the exact form they assume is no longer a matter of specific genetic control. The forest ape that became a ground ape that became a hunting ape that became a territorial ape has become a cultural ape, and we must call a temporary halt.

It is worth re-iterating here that, in this book, we are not concerned with the massive cultural explosions that followed, of which the naked ape of today is so proud—the dramatic progression that led him, in a mere half-million years, from making a fire to making a space-craft. It is an exciting story, but the naked ape is in danger of being dazzled by it all and forgetting that beneath the surface gloss he is still very much a primate. ('An ape's an ape, a varlet's a varlet, though they be clad in silk or scarlet.') Even a space ape must urinate.

Only by taking a hard look at the way in which we have originated and then by studying the biological aspects of the way we behave as a species today, can we really acquire a balanced, objective understanding of our extraordinary existence.

If we accept the history of our evolution as it has been out-lined here, then one fact stands out clearly: namely, that we have arisen essentially as primate predators. Amongst existing monkeys and apes, this makes us unique, but major conversions of this kind are not unknown in other groups. The giant panda, for instance, is a perfect case of the reverse process. Whereas we are vegetarians turned carnivores, the panda is a carnivore turned vegetarian, and like us it is in many ways an extraordinary and unique creature. The point is that a major switch of this sort produces an animal with a dual personality. Once over the threshold, it plunges into its new role with great evolutionary energy—so much so that it carries with it many of its old traits. Insufficient time has passed for it to throw off all its old characteristics while it is hurriedly donning the new ones. When the ancient fishes first conquered dry land, their new terrestrial qualities raced ahead while they continued to drag their old watery ones with them. It takes millions of years to perfect a dramatically new animal model, and the pioneer forms are usually very odd mixtures indeed. The naked ape is such a mixture. His whole body, his way of life, was geared to a forest existence, and then suddenly (suddenly in evolutionary terms) he was jettisoned into a world where he could survive only if he began to live like a brainy, weapon-toting wolf. We must examine now exactly how this affected not only his body, but especially his behaviour, and in what form we experience the influence of this legacy at the present day.

One way of doing this is to compare the structure and the way of life of a 'pure' fruit-picking primate with a 'pure' carnivore. Once we have cleared our minds about the essential differences that relate to their two contrasted methods of feeding, we can then re-examine the naked ape situation to see how the mixture has been worked out.

The brightest stars in the carnivore galaxy are, on the one hand, the wild dogs and wolves, and, on the other, the big cats such as the lions, tigers and leopards. They are beautifully equipped with delicately perfected sense organs. Their sense of hearing is acute, and their external ears can twist this way and that to pick up the slightest rustle or snort. Their eyes, although poor on static detail and colour, are incredibly responsive to the tiniest movement. Their sense of smell is so good that it is difficult for us to comprehend it. They must be able to experience a virtual landscape of odours. Not only are they capable of detecting an individual smell with unerring precision, but they are also able to pick out the separate component odours of a complex smell. Experiments carried out with dogs in 1953 indicated that their sense of smell was between a million and a thousand million times as accurate as ours. These astonishing results have since been queried, and later, more careful tests have not been able to confirm them, but even the most cautious estimates put the dog's sense of smell at about a hundred times better than ours.

In addition to this first-rate sensory equipment, the wild

dogs and big cats have a wonderfully athletic physique. The cats have specialized as lightning sprinters, the dogs as long-distance runners of great stamina. At the kill they can bring into action powerful jaws, sharp, savage teeth and, in the case of the big cats, massively muscular front limbs armed with huge, dagger-pointed claws.

For these animals, the act of killing has become a goal in itself, a consummatory act. It is true that they seldom kill wantonly or wastefully, but if, in captivity, one of these carnivores is given ready-killed food, its urge to hunt is far from satisfied. Every time a domestic dog is taken for a walk by its master, or has a stick thrown for it to chase and catch, it is having its basic need to hunt catered for in a way that no amount of canned dog-food will subdue. Even the most over-stuffed domestic cat demands a nocturnal prowl and the chance to leap on an unsuspecting bird.

Their digestive system is geared to accept comparatively long periods of fasting followed by bloating gorges. (A wolf, for instance, can eat one-fifth of its total body weight at one meal—the equivalent of you or me devouring a 30–40 lb. steak at a single sitting.) Their food is of high nutritional value and there is little wastage. Their faeces, however, are messy and smelly and defecation involves special behaviour patterns. In some cases the faeces are actually buried and the site carefully covered over. In others, the act of defecating is always carried out at a considerable distance from the home base. When young cubs foul the den, the faeces are eaten by the mother and the home is kept clean in this way.

Simple food storage is undertaken. Carcasses, or parts of them, may be buried, as with dogs and certain kinds of cats; or they may be carried up into a tree-larder, as with the leopard. The periods of intensive athletic activity during the hunting and killing phases are interspersed with periods of great laziness and relaxation. During social encounters the savage weapons so vital to the kill constitute a potential threat to life and limb in any minor disputes and rivalries. If two wolves or two lions fall out, they are both so heavily armed that fighting could easily, in a matter of seconds, lead to mutilation or death. This could seriously endanger the survival of the species and during the long course of the evolution that gave these species their lethal prey-killing weapons, they have of necessity also developed powerful inhibitions about using their weapons on other members of their own species. These inhibitions appear to have a specific genetic basis: they do not have to be learned. Special submissive postures have been evolved which automatically appease a dominant animal and inhibit its attack. The possession of these signals is a vital part of the way of life of the 'pure' carnivores.

The actual method of hunting varies from species to species. In the leopard it is a matter of solitary stalking or hiding, and a last-minute pounce. For the cheetah it is a careful prowl followed by an all-out sprint. For the lion it is usually a group action, with the prey driven in panic by one lion towards others in hiding. For a pack of wolves it may involve an encircling manoeuvre followed by a group kill. For a pack of African hunting dogs it is typically a ruthless drive, with one dog after another going in to the attack until the fleeing prey is weakened from loss of blood.

Recent studies in Africa have revealed that the spotted hyaena is also a savage pack-hunter and not, as has always been thought, primarily a scavenger. The mistake has been made because hyaena packs form only at night and minor scavenging has always been recorded during the day. When dusk falls, the hyaena becomes a ruthless killer, just as efficient as the hunting dog is during the day. Up to thirty animals may hunt together. They easily out-pace the zebras or antelopes they are pursuing, which dare not travel at their full day-time speeds. The hyaenas start tearing at the legs of any prey in reach until one is sufficiently wounded to fall back from the fleeing herd. All the hyaenas then converge on this one, tearing out its soft parts until it drops and is killed. Hyaenas base themselves at communal den-sites. The group or 'clan' using this home base may number between ten and a hundred. Females stick closely to the area around this base, but the males are more mobile and may wander off into other regions. There is considerable aggression between clans if wandering individuals are caught off their own clan territory, but there is little aggression between the members of any one clan.

Food-sharing is known to be practised in a number of species. Of course, at a large kill there is meat enough for the whole hunting group and there need be little squabbling, but in some instances the sharing is taken further than that. African hunting dogs, for instance, are known to re-gurgitate food to one another after a hunt is over. In some cases they have done this to such an extent that they have been referred to as having a 'communal stomach'.

Carnivores with young go to considerable trouble to pro-vide food for their growing offspring. Lionesses will hunt and carry meat back to the den, or they will swallow large hunks of it and then re-gurgitate it for the cubs. Male lions have occasionally been reported to assist in this matter, but it does not appear to be a common practice. Male wolves, on the other hand, have been known to travel up to fifteen miles to obtain food for both the female and her young. Large meaty bones may be carried back for the young to gnaw, or hunks of meat may be swallowed at the kill and then re-gurgitated at the entrance to the den.

These, then, are some of the main features of the specialist carnivores, as they relate to their hunting way of life. How do they compare with those of the typical fruit-picking monkeys and apes?

The sensory equipment of the higher primates is much more dominated by the sense of vision than the sense of smell. In their tree-climbing world, seeing well is far more important than smelling well, and the snout has shrunk considerably, giving the eyes a much better view. In searching for food, the colours of fruits are helpful clues, and, unlike the carnivores, primates have evolved good colour vision. Their eyes are also better at picking out static details. Their food is static, and detecting minute movements is less vital than recognizing subtle differences in shape and texture. Hearing is important, but less so than for the tracking killers, and their external ears are smaller and lack the twisting mobility of those of the carnivores. The sense of taste is more refined. The diet is more varied and highly flavoured – there is more to taste. In particular there is a strong positive response to sweet-tasting objects.

The primate physique is good for climbing and clambering, but it is not built for high-speed sprinting on the ground, or for lengthy endurance feats. This is the agile body of an acrobat rather than the burly frame of a powerful athlete. The hands are good for grasping, but not for tearing or striking. The jaws and teeth are reasonably strong, but nothing like the massive, clamping, crunching apparatus of the carnivores. The occasional killing of small, insignificant prey requires no gargantuan efforts. Killing is not, in fact, a basic part of the primate way of life.

Feeding is spread out through much of the day. Instead of great gorging feasts followed by long fasts, the monkeys and apes keep on and on munching – a life of non-stop snacks. There are, of course, periods of rest, typically in the middle of the day and during the night, but the contrast is nevertheless a striking one. The static food is always there, just waiting to be plucked and eaten. All that is necessary is for the animals to move from one feeding-place to another, as their tastes change, or as the fruits come in and out of season. No food storage takes place except, in a very temporary way, in the bulging cheek pouches of certain monkeys.

The faeces are less smelly than those of the meat-eaters and no special behaviour has developed to deal with their disposal, since they drop down out of the trees and away from the animals. As the group is always on the move, there is little danger of a particular area becoming unduly fouled or smelly. Even the great apes that bed down in special sleeping-nests make a new bed at a new site each night, so that there is little need to worry about nest sanitation. (All the same, it is rather surprising to discover that 99 per cent of abandoned gorilla nests in one area of Africa had gorilla dung inside them, and that in 73 per cent the animals had actually been lying in it. This is bound to constitute a disease risk by increasing the chances of re-infection, and is a remarkable illustration of the basic faecal disinterest of primates.)

Because of the static nature and abundance of the food, there is no need for the primate group to split up to search for it. They can move, flee, rest and sleep together in a close-knit community, with every member keeping an eye on the movements and actions of every other. Each individual of the group will at any one moment have a reasonably good idea of what everyone else is doing. This is a very non-carnivore procedure. Even in those species of primates that do split up from time to time, the smaller unit is never composed of a single individual. A solitary monkey or ape is a vulnerable creature. It lacks the powerful natural weapons of the carnivore and in isolation falls easy prey to the stalking killers.

The co-operative spirit that is present in such pack-hunters as wolves is largely absent from the world of the primate. Competitiveness and dominance is the order of his day. Competition in the social hierarchy is, of course, present in both groups, but it is less tempered by co-operative action in the case of monkeys and apes. Complicated, co-ordinated manœuvres are also unnecessary: sequences of feeding action do not need to be strung together in such a complex way. The primate can live much more from minute to minute, from hand to mouth.

Because the primate's food supply is all around it for the taking, there is little need to cover great distances. Groups of wild gorillas, the largest of the living primates, have been carefully studied and their movements traced, so that we now know that they travel on the average about a third of a mile a day. Sometimes they move only a few hundred feet. Carnivores, by contrast, must frequently travel many miles on a single hunting trip. In some instances they have been known to travel over fifty miles on a hunting journey, taking several days before returning to their home base. This act of returning to a fixed home base is typical of the carnivores, but is far less common amongst the monkeys and apes. True, a group of primates will live in a reasonably clearly defined home range, but at night it will probably bed down wherever it happens to have ended up in its day's meanderings. It will get to know the general region in which it lives because it is always wandering back and forth across it, but it will tend to use the whole area in a much more haphazard way. Also, the interaction between one troop and the next will be less defensive and less aggressive than is the case with carnivores. A territory is, by definition, a defended area, and primates are not therefore, typically, territorial animals.

A small point, but one that is relevant here, is that carnivores have fleas but primates do not. Monkeys and apes are plagued by lice and certain other external parasites but, contrary to popular opinion, they are completely flealess, for one very good reason. To understand this, it is necessary to examine the life-cycle of the flea. This insect lays its eggs, not on the body of its host, but amongst the detritus of its victim's sleeping quarters. The eggs take three days to hatch into small, crawling maggots. These larvae do not feed on blood, but on the waste matter that has accumulated in the dirt of the den or lair. After two weeks they spin a cocoon and pupate. They remain in this dormant condition for approximately two more weeks before emerging as adults, ready to hop on to a suitable host body. So for at least the first month of its life a flea is cut off from its host species. It is clear from this why a nomadic mammal, such as a monkey or ape, is not troubled by fleas. Even if a few stray fleas do happen on to one and mate successfully, their eggs will be left behind as the primate group moves on, and when the pupae hatch there will be no host 'at home' to continue the relationship. Fleas are therefore parasites only of animals with a fixed home base, such as the typical carnivores. The significance of this point will become clear in a moment.

In contrasting the different ways of life of the carnivores and the primates, I have naturally concentrated on the typical open-country hunters on the one hand, and the typical forest-dwelling fruit-pickers on the other. There are certain minor exceptions to the general rules on both sides, but we must concentrate now on the one major exception—the naked ape. To what extent was he able to modify himself, to blend his frugivorous heritage with his newly adopted carnivory? Exactly what kind of an animal did this cause him to become?

To start with, he had the wrong kind of sensory equipment for life on the ground. His nose was too weak and his ears not sharp enough. His physique was hopelessly inadequate for arduous endurance tests and for lightning sprints. In personality he was more competitive than co-operative and no doubt poor on planning and concentration. But fortunately he had an excellent brain, already better in terms of general intelligence than that of his carnivore rivals. By bringing his body up into a vertical position, modifying his hands in one way and his feet in another, and by improving his brain still further and using it as hard as he could, he stood a chance.

This is easy to say, but it took a long time to do, and it had all kinds of repercussions on other aspects of his daily life, as we shall see in later chapters. All we need concern ourselves with for the moment is how it was achieved and how it affected his hunting and feeding behaviour.

As the battle was to be won by brain rather than brawn, some kind of dramatic evolutionary step had to be taken to greatly increase his brain-power. What happened was rather odd: the hunting ape became an infantile ape. This evolutionary trick is not unique; it has happened in a number of quite separate cases. Put very simply, it is a process (called neoteny) by which certain juvenile or infantile characters are retained and prolonged into adult life. (A famous example is the axolotl, a kind of salamander that may remain a tadpole all its life and is capable of breeding in this condition.)

The way in which this process of neoteny helps the primate brain to grow and develop is best understood if we consider the unborn infant of a typical monkey. Before birth the brain of the monkey foetus increases rapidly in size and complexity. When the animal is born its brain has already attained seventy per cent of its final adult size. The remaining thirty per cent of growth is quickly completed in the first six months of life. Even a young chimpanzee completes its brain-growth within twelve months after birth. Our own species, by contrast, has at birth a brain which is only twenty-three per cent of its final adult size. Rapid growth continues for a further six years after birth, and the whole growing process is not complete until about the twenty-third year of life.

For you and me, then, brain-growth continues for about ten years *after* we have attained sexual maturity, but for the chimpanzee it is completed six or seven years *before* the animal becomes reproductively active. This explains very clearly what is meant by saying that we became infantile apes, but it is essential to qualify this statement. We (or rather, our hunting ape ancestors) became infantile in certain ways, but not in others. The rates of development of our various properties got out of phase. While our reproductive systems raced ahead, our brain-growth dawdled behind. And so it was with various other parts of our make-up, some being greatly slowed down, others a little, and still others not at all. In other words, there was a process of differential infantilism. Once the trend was under way, natural selection would favour the slowing down of any parts of the animal's make-up that helped it to survive in its hostile and difficult new environment. The brain was not the only part of the body affected: the body posture was also influenced in the same way. An unborn mammal has the axis of its head at right angles to the axis of its trunk. If it were born in this condition its head would point down at the ground as it moved along on all fours, but before birth occurs the head rotates backwards so that its axis is in line with that of the trunk. Then, when it is born and walking along, its head points forwards in the approved manner. If such an animal began to walk along on its hind legs in a vertical posture, its head would point upwards, looking at the sky. For a vertical animal, like the hunting ape, it is important therefore to retain the foetal angle of the head, keeping it at right angles to the body so that, despite the new locomotion position, the head faces forwards. This is, of course, what has happened and, once again, it is an example of neoteny, the pre-birth stage being retained into the post-birth and adult life.

Many of the other special physical characters of the hunting ape can be accounted for in this way: the long slender neck, the flatness of the face, the small size of the teeth and their late eruption, the absence of heavy brow ridges and the non-rotation of the big toe.

The fact that so many separate embryonic characteristics were potentially valuable to the hunting ape in his new role was the evolutionary breakthrough that he needed. In one neotenous stroke he was able to acquire both the brain he needed and the body to go with it. He could run vertically with his hands free to wield weapons, and at the same time he developed the brain that could develop the weapons. More than that, he not only became brainier at manipulating objects, but he also had a longer childhood during which he could learn from his parents and other adults. Infant monkeys and chimpanzees are playful, exploratory and inventive, but this phase dies quickly. The naked ape's infancy was, in these respects, extended right through into his sexually adult life. There was plenty of time to imitate and learn the special techniques that had been devised by previous generations. His weaknesses as a physical and instinctive hunter could be more than compensated for by his intelligence and his imitative abilities. He could be taught by his parents as no animal had ever been taught before.

But teaching alone was not enough. Genetic assistance was required. Basic biological changes in the nature of the hunting ape had to accompany this process. If one simply took a typical, forest-living, fruit-picking primate of the kind described earlier, and gave it a big brain and a hunting body, it would be difficult for it to become a successful hunting ape without some other modifications. Its basic behaviour patterns would be wrong. It might be able to think things out and plan in a very clever way, but its more fundamental animal urges would be of the wrong type. The teaching would be working *against* its natural tendencies, not only in its feeding behaviour, but also in its general social, aggressive and sexual behaviour, and in all the other basic behavioural aspects of its earlier primate existence. If genetically controlled changes were not wrought here too, then the new education of the young hunting ape would be an impossibly uphill task. Cultural training can achieve a great deal, but no matter how brilliant the machinery of the higher centres of the brain, it needs a considerable degree of support from the lower regions.

If we look back now at the differences between the typical 'pure' carnivore and the typical 'pure' primate, we can see how this probably came about. The advanced carnivore separates the actions of food-seeking (hunting and killing) from the actions of eating. They have become two distinct motivational systems with only partial dependence one on the other. This has come about because the whole sequence is so lengthy and arduous. The act of feeding is too remote, and so the action of killing has to become a reward in itself. Researches with cats have even indicated that the sequence there has become further sub-divided. Catching the prey, killing it, preparing it (plucking it), and eating it, each have their own partially independent motivational systems. If one of these patterns of behaviour is satiated, it does not automatically satiate the others.

For the fruit-picking primate the situation is entirely different. Each feeding sequence, comprising simple food-searching and then immediate eating, is comparatively so brief that no splitting up into separate motivational systems is necessary. This is something that would have to be changed, and changed radically, in the case of the hunting ape. Hunting would have to bring its own reward, it could no longer simply act as an appetitive sequence leading up to the consummatory meal. Perhaps, as in the cat, hunting, killing and preparing the food would each develop their own separate, independent goals, would each become ends in themselves. Each would then have to find expression and one could not be damped down by satisfying another. If we examine—as we shall be doing in a later chapter—the feeding behaviour of present-day naked apes, we shall see that there are plenty of indications that something like this did occur.

In addition to becoming a biological (as opposed to a cultural) killer, the hunting ape also had to modify the timing arrangements of his eating behaviour. Minute-by-minute snacks were out and big, spaced meals were in. Food storage was practised. A basic tendency to return to a fixed home base had to be built in to the behavioural system. Orientation and homing abilities had to be improved. Defecation had to become a spatially organized pattern of behaviour, a private (carnivore) activity instead of a communal (primate) one.

I mentioned earlier that one outcome of using a fixed home base is that it makes parasitization by fleas possible. I also said that carnivores have fleas, but primates do not. If the hunting ape was unique amongst primates in having a fixed base, then we would also expect him to break the primate rule concerning fleas, and this certainly seems to be the case. We know that today our species is parasitized by these insects and that we have our own special kind of flea—one that belongs to a different species from other fleas, one that has evolved with us. If it had sufficient time to develop into a new species, then it must have been with us for a very long while indeed, long enough to have been an unwelcome companion right back in our earliest hunting-ape days.

Socially the hunting ape had to increase his urge to communicate and to co-operate with his fellows. Facial expressions and vocalizations had to become more complicated. With the new weapons to hand, he had to develop powerful signals that would inhibit attacks within the social group. On the other hand, with a fixed home base to defend, he had to develop stronger aggressive responses to members of rival groups.

Because of the demands of his new way of life, he had to reduce his powerful primate urge never to leave the main body of the group.

As part of his new-found co-operativeness and because of the erratic nature of the food supply, he had to begin to share out his food. Like the paternal wolves mentioned earlier, the hunting ape males also had to carry food supplies home for the nursing females and their slowly growing young. Paternal behaviour of this kind had to be a new development, for the general primate rule is that virtually all parental care comes from the mother. (It is only a wise primate, like our hunting ape, that knows its own father.)

Because of the extremely long period of dependency of the young and the heavy demands made by them, the females found themselves almost perpetually confined to the home base. In this respect the hunting ape's new way of life threw up a special problem, one that it did not share with the typical 'pure' carnivores: the role of the sexes had to become more distinct. The hunting parties, unlike those of the 'pure' carnivores, had to become all-male groups. If anything was going to go against the primate grain, it was this. For a virile primate male to go off on a feeding trip and leave his females unprotected from the advances of any other males that might happen to come by, was unheard of. No amount of cultural training could put this right. This was something that demanded a major shift in social behaviour.

The answer was the development of a pair-bond. Male and female hunting apes had to fall in love and remain faithful to one another. This is a common tendency in many other groups of animals, but is rare amongst primates. It solved three problems in one stroke. It meant that the females remained bonded to their individual males and faithful to them while they were away on the hunt. It meant that serious sexual rivalries between the males were reduced. This aided their developing co-operativeness. If they were to hunt together successfully, the weaker males as well as the stronger ones had to play their part. They had to play a central role and could not be thrust to the periphery of society, as happens in so many primate species. What is more, with his newly developed and deadly artificial weapons, the hunting ape male was under strong pressure to reduce any source of disharmony within the tribe. Thirdly, the development of a one-male-one-female breeding unit meant that the offspring also benefited. The heavy task of rearing and training the slowly developing young demanded a cohesive family unit. In other groups of animals, whether they are fishes, birds or mammals, when there is too big a burden for one parent to bear alone, we see the development of a powerful pair-bond, tying the male and female parents together throughout the breeding season. This, too, is what occurred in the case of the hunting ape.

In this way, the females were sure of their males' support and were able to devote themselves to their maternal duties. The males were sure of their females' loyalty, were prepared to leave them for hunting, and avoided fighting over them. And the offspring were provided with the maximum of care and attention. This certainly sounds like an ideal solution, but it involved a major change in primate socio-sexual behaviour and, as we shall see later, the process was never really perfected. It is clear from the behaviour of our species today that the trend was only partially completed and that our earlier primate urges keep on re-appearing in minor forms.

This is the manner, then, in which the hunting ape took on the role of a lethal carnivore and changed his primate ways accordingly. I have suggested that they were basic biological changes rather than mere cultural ones, and that the new species changed genetically in this way. You may consider this an unjustified assumption. You may feel — such is the power of cultural indoctrination — that the modifications could easily have been made by training and the development of new traditions. I doubt this. One only has to look at the behaviour of our species at the present day to see that this is not so. Cultural developments have given us more and more impressive technological advances, but wherever these clash with our basic biological properties they meet strong resistance. The fundamental patterns of behaviour laid down in our early days as hunting apes still shine through all our affairs, no matter how lofty they may be. If the organization of our earthier activities — our feeding, our fear, our aggression, our sex, our parental care — had been developed solely by cultural means, there can be little doubt that we would have got it under better control by now, and twisted it this way and that to suit the increasingly extraordinary demands put upon it by our technological advances. But we have not done so. We have repeatedly bowed our heads before our animal nature and tacitly admitted the existence of the complex beast that stirs within us. If we are honest, we will confess that it will take millions of years, and the same genetic process of natural selection that put it there, to change it. In the meantime, our unbelievably complicated civilizations will be able to prosper only if we design them in such a way that they do not clash with or tend to suppress our basic animal demands. Unfortunately our thinking brain is not always in harmony with our feeling brain. There are many examples showing where things have gone astray, and human societies have crashed or become stultified.

In the chapters that follow we will try to see how this has happened, but first there is one question that must be answered —the question that was asked at the beginning of this chapter. When we first encountered this strange species we noted that it had one feature that stood out immediately from the rest, when it was placed as a specimen in a long row of primates. This feature was its naked skin, which led me as a zoologist to name the creature 'the naked ape'. We have since seen that it could have been given any number of suitable names: the vertical ape, the tool-making ape, the brainy ape, the territorial ape, and so on. But these were not the first things we noticed. Regarded simply as a zoological specimen in a museum, it is the nakedness that has the immediate impact, and this is the name we will stick to, if only to bring it into line with other zoological studies and remind us that this is the special way in which we are approaching it. But what is the significance of this strange feature? Why on earth should the hunting ape have become a naked ape?

Unfortunately fossils cannot help us when it comes to differences in skin and hair, so that we have no idea as to exactly when the great denudation took place. We can be fairly certain that it did not happen before our ancestors left their forest homes. It is such an odd development that it seems much more likely to have been yet another feature of the great transformation scene on the open plains. But exactly how did it occur, and how did it help the emerging ape to survive?

This problem has puzzled experts for a long time and many imaginative theories have been put forward. One of the most promising ideas is that it was part and parcel of the process of neoteny. If you examine an infant chimpanzee at birth you will find that it has a good head of hair, but that its body is almost naked. If this condition was delayed into the animal's adult life by neoteny, the adult chimpanzee's hair condition would be very much like ours.

It is interesting that in our own species this neotenous suppression of hair growth has not been entirely perfected. The growing foetus starts off on the road towards typical mammalian hairiness, so that between the sixth and eighth months of its life in the womb it becomes almost completely covered in a fine hairy down. This foetal coat is referred to as the lanugo and it is not shed until just before birth. Premature babies sometimes enter the world still wearing their lanugo, much to the horror of their parents, but, except in very rare cases, it soon drops away. There are no more than about thirty recorded instances of families producing offspring that grow up to be fully furred adults.

Even so, all adult members of our species do have a large number of body hairs — more, in fact, than our relatives the chimpanzees. It is not so much that we have lost whole hairs as that we have sprouted only puny ones. (This does not, incidentally, apply to all races — negroes have undergone a real as well as an apparent hair loss.) This fact has led certain anatomists to declare that we cannot consider ourselves as a hairless or naked species, and one famous authority went so far as to say that the statement that we are 'the least hairy of all the primates is, therefore, very far from being true; and the numerous quaint theories that have been put forward to account for the imagined loss of hairs are, mercifully, not needed.' This is clearly nonsensical. It is like saying that because a blind man has a pair of eyes he is not blind. Functionally, we are stark naked and our skin is fully exposed to the outside world. This state of affairs still has to be explained, regardless of how many tiny hairs we can count under a magnifying lens.

The neoteny explanation only gives a clue as to how the nakedness could have come about. It does not tell us anything about the value of nudity as a new character that helped the naked ape to survive better in his hostile environment. It might be argued that it had no value, that it was merely a by-product of other, more vital neotenous changes, such as the brain development. But as we have already seen, the process of neoteny is one of differential retarding of developmental processes. Some things slow down more than others—the rates of growth get out of phase. It is hardly likely, therefore, that an infantile trait as potentially dangerous as nakedness was going to be allowed to persist simply because other changes were slowing down. Unless it had some special value to the new species, it would be quickly dealt with by natural selection.

What, then, was the survival value of naked skin? One explanation is that when the hunting ape abandoned its nomadic past and settled down at fixed home bases, its dens became heavily infested with skin parasites. The use of the same sleeping places night after night is thought to have provided abnormally rich breeding-grounds for a variety of ticks, mites, fleas and bugs, to a point where the situation provided a severe disease risk. By casting off his hairy coat, the den-dweller was better able to cope with the problem.

There may be an element of truth in this idea, but it can hardly have been of major importance. Few other den-dwelling mammals—and there are hundreds of species to pick from—have taken this step. Nevertheless, if nakedness was developed in some other connection, it might make it easier to remove troublesome skin parasites, a task which today still occupies a great deal of time for the hairier primates.

Another thought along similar lines is that the hunting ape had such messy feeding habits that a furry coat would soon become clogged and messy and, again, a disease risk. It is pointed out that vultures, who plunge their heads and necks into gory carcasses, have lost the feathers from these members; and that the same development, extended over the whole body, may have occurred among the hunting apes. But the ability to develop tools to kill and skin the prey can hardly have preceded the ability to use other objects to clean the hunters' hair. Even a chimpanzee in the wild will occasionally use leaves as toilet paper when in difficulties with defecation.

A suggestion has even been put forward that it was the development of fire that led to the loss of the hairy coat. It is argued that the hunting ape will have felt cold only at night and that, once he had the luxury of sitting round a camp fire, he was able to dispense with his fur and thus leave himself in a better state for dealing with the heat of the day.

Another, more ingenious theory is that, before he became a hunting ape, the original ground ape that had left the forests went through a long phase as an aquatic ape. He is envisaged as moving to the tropical sea-shores in search of food. There he will have found shellfish and other sea-shore creatures in comparative abundance, a food supply much richer and more attractive than that on the open plains. At first he will have groped around in the rock pools and the shallow water, but gradually he will have started to swim out to greater depths and dive for food. During this process, it is argued, he will have lost his hair like other mammals that have returned to the sea. Only his head, protruding from the surface of the water, would retain the hairy coat to protect him from the direct glare of the sun. Then, later on, when his tools (originally developed for cracking open shells) became sufficiently advanced, he will have spread away from the cradle of the sea-shore and out into the open land spaces as an emerging hunter.

It is held that this theory explains why we are so nimble in the water today, while our closest living relatives, the chimpanzees, are so helpless and quickly drown. It explains our streamlined bodies and even our vertical posture, the latter supposedly having developed as we waded into deeper and deeper water. It clears up a strange feature of our body-hair tracts. Close examination reveals that on our backs the directions of our tiny remnant hairs differ strikingly from those of other apes. In us they point diagonally backwards and inwards towards the spine. This follows the direction of the flow of water passing over a swimming body and indicates that, if the coat of hair was modified before it was lost, then it was modified in exactly the right way to reduce resistance when swimming. It is also pointed out that we are unique amongst all the primates in being the only one to possess a thick layer of sub-cutaneous fat. This is interpreted as the equivalent of the blubber of a whale or seal, a compensatory insulating device. It is stressed that no other explanation has been given for this feature of our anatomy. Even the sensitive nature of our hands is brought into play on the side of the aquatic theory. A reasonably crude hand can, after all, hold a stick or a rock, but it takes a subtle, sensitized hand to feel for food in the water. Perhaps this was the way that the ground ape originally acquired its super-hand, and then passed it on ready-made to the hunting ape. Finally, the aquatic theory needles the traditional fossil-hunters by pointing out that they have been singularly unsuccessful in unearthing the vital missing links in our ancient past, and gives them the hot tip that if they would only take the trouble to search around the areas that constituted the African coastal sea-shores of a million or so years ago, they might find something that would be to their advantage.

Unfortunately this has yet to be done and, despite its most appealing indirect evidence, the aquatic theory lacks solid support. It neatly accounts for a number of special features, but it demands in exchange the acceptance of a hypothetical major evolutionary phase for which there is no direct evidence. (Even if eventually it does turn out to be true, it will not clash seriously with the general picture of the hunting ape's evolution out of a ground ape. It will simply mean that the ground ape went through a rather salutary christening ceremony.)

An argument along entirely different lines has suggested that, instead of developing as a response to the physical environment, the loss of hair was a social trend. In other words it arose, not as a mechanical device, but as a signal. Naked patches of skin can be seen in a number of primate species and in certain instances they appear to act as species recognition marks, enabling one monkey or ape to identify another as belonging to its own kind, or some other. The loss of hair on the part of the hunting ape is regarded simply as an arbitrarily selected characteristic that happened to be adopted as an identity badge by this species. It is of course undeniable that stark nudity must have rendered the naked ape startlingly easy to identify, but there are plenty of other less drastic ways of achieving the same end, without sacrificing a valuable insulating coat.

Another suggestion along the same lines pictures the loss of hair as an extension of sexual signalling. It is claimed that male mammals are generally hairier than their females and that, by extending this sex difference, the female naked ape was able to become more and more sexually attractive to the male. The trend to loss of hair would affect the male, too, but to a lesser extent and with special areas of contrast, such as the beard.

This last idea may well explain the sex differences as regards hairiness but, again, the loss of body insulation would be a high price to pay for a sexy appearance alone, even with sub-cutaneous fat as a partial compensating device. A modification of this idea is that it was not so much the appearance as the sensitivity to touch that was sexually important. It can be argued that by exposing their naked skins to one another during sexual encounters, both male and female would become more highly sensitized to erotic stimuli. In a species where pair-bonding was evolving, this would heighten the excite-ment of sexual activities and would tighten the bond between the pair by intensifying copulatory rewards.

Perhaps the most commonly held explanation of the hairless condition is that it evolved as a cooling device. By coming out of the shady forests the hunting ape was exposing himself to much greater temperatures than he had previously experienced, and it is assumed that he took off his hairy coat to prevent himself from becoming over-heated. Superficially this is reasonable enough. We do, after all, take our jackets off on a hot summer's day. But it does not stand up to closer scrutiny. In the first place, none of the other animals (of roughly our size) on the open plains have taken this step. If it was as simple as this we might expect to see some naked lions and naked jackals. Instead they have short but dense coats. Exposure of the naked skin to the air certainly increases the chances of heat loss, but it also increases heat gain at the same time and risks damage from the sun's rays, as any sun-bather will know. Experiments in the desert have shown that the wearing of light clothing may reduce heat loss by curtailing water evaporation, but it also reduces heat gain from the environ-ment to 55 per cent of the figure obtained in a state of total nudity. At really high temperatures, heavier, looser clothing of the type favoured in Arab countries is a better protection than even light clothing. It cuts down the in-coming heat, but at the same time allows air to circulate around the body and aid in the evaporation of cooling sweat.

Clearly the situation is more complicated than it at first appears. A great deal will depend on the exact temperature levels of the environment and on the amount of direct sun-shine. Even if we suppose that the climate was suitable for hair loss—that is, moderately hot, but not intensely hot—we still have to explain the striking difference in coat condition between the naked ape and the other open-country carni-vores.

There is one way we can do this, and it may give the best answer yet to the whole problem of our nakedness. The essential difference between the hunting ape and his carnivore rivals was that he was not physically equipped to make light-rting dashes after his prey or even to undertake long endurance pursuits. But this is nevertheless precisely what he had to do. He succeeded because of his better brain, leading to more intelligent manœuvring and more lethal weapons, but despite this such efforts must have put a huge strain on him in simple physical terms. The chase was so important to him that he would have to put up with this, but in the process he must have experienced considerable over-heating. There would be a strong selection pressure working to reduce this over-heating and any slight improvement would be favoured, even if it meant sacrifices in other directions. His very survival depended on it. This surely was the key factor operating in the conversion of a hairy hunting ape into a naked ape. With neoteny to help the process on its way, and with the added advantages of the minor secondary benefits already men-tioned, it would become a viable proposition. By losing the heavy coat of hair and by increasing the number of sweat glands all over the body surface, considerable cooling could be achieved—not for minute-by-minute living, but for the supreme moments of the chase—with the production of a generous film of evaporating liquid over his air-exposed, straining limbs and trunk.

This system would not succeed, of course, if the climate were too intensely hot, because of damage to the exposed skin, but in a moderately hot environment it would be acceptable. It is interesting that the trend was accompanied by the development of a sub-cutaneous fat layer, which indicates that there was a need to keep the body warm at other times. If this appears to counterbalance the loss of the hairy coat, it should be remembered that the fat layer helps to retain the body heat in cold conditions, without hindering the evaporation of sweat when over-heating takes place. The combination of reduced hair, increased sweat glands, and the fatty layer under the skin appears to have given our hard-working ancestors just what they needed, bearing in mind that hunting was one of the most important aspects of their new way of life.

So there he stands, our vertical, hunting, weapon-toting, territorial, neotenous, brainy, Naked Ape, a primate by ancestry and a carnivore by adoption, ready to conquer the world. But he is a very new and experimental departure, and new models frequently have imperfections. For him the main troubles will stem from the fact that his culturally operated advances will race ahead of any further genetic ones. His genes will lag behind, and he will be constantly reminded that, for all his environment-moulding achievements, he is still at heart a very naked ape.

At this point we can leave his past behind us and see how we find him faring today. How *does* the modern naked ape behave? How does he tackle the age-old problems of feeding, fighting, mating, and rearing his young? How much has his computer of a brain been able to reorganize his mammalian urges? Perhaps he has had to make more concessions than he likes to admit. We shall see.

Chapter Two

*

SEX

Sexually the naked ape finds himself today in a somewhat confusing situation. As a primate he is pulled one way, as a carnivore by adoption he is pulled another, and as a member of an elaborate civilized community he is pulled yet another.

To start with, he owes all his basic sexual qualities to his fruit-picking, forest-ape ancestors. These characteristics were then drastically modified to fit in with his open-country, hunting way of life. This was difficult enough, but then they, in turn, had to be adapted to match the rapid development of an increasingly complex and culturally determined social structure.

The first of these changes, from a sexual fruit-picker to a sexual hunter, was achieved over a comparatively long period of time and with reasonable success. The second change has been less successful. It has happened too quickly and has been forced to depend upon intelligence and the application of learned restraint rather than on biological modifications based on natural selection. It could be said that the advance of civilization has not so much moulded modern sexual behaviour, as that sexual behaviour has moulded the shape of civilization. If this seems to be a rather sweeping statement, let let me first put my case and then we can return to the argument at the end of the chapter.

To begin with we must establish precisely how the naked

ape does behave today when indulging in sexual behaviour. This is not as easy as it sounds, because of the great variability that exists, both between and within societies. The only solution is to take average results from large samples of the most successful societies. The small, backward, and unsuccessful societies can largely be ignored. They may have fascinating and bizarre sexual customs, but biologically speaking they no longer represent the mainstream of evolution. Indeed, it may very well be that their unusual sexual behaviour has helped to turn them into biological failures as social groups.

Most of the detailed information we have available stems from a number of painstaking studies carried out in recent years in North America and based largely on that culture. Fortunately it is biologically a very large and successful culture and can, without undue fear of distortion, be taken as representative of the modern naked ape.

Sexual behaviour in our species goes through three characteristic phases: pair-formation, pre-copulatory activity, and copulation, usually but not always in that order. The pair-formation stage, usually referred to as courtship, is remarkably prolonged by animal standards, frequently lasting for weeks or even months. As with many other species it is characterized by tentative, ambivalent behaviour involving conflicts between fear, aggression and sexual attraction. The nervousness and hesitancy is slowly reduced if the mutual sexual signals are strong enough. These involve complex facial expressions, body postures and vocalizations. The latter involve the highly specialized and symbolized sound signals of speech, but equally importantly they present to the member of the opposite sex a distinctive vocalization tone. A courting couple is often referred to as 'murmuring sweet nothings' and this phrase sums up clearly the significance of the tone of voice as opposed to what is being spoken.

After the initial stages of visual and vocal display, simple body contacts are made. These usually accompany locomotion, which is now considerably increased when the pair are together. Hand-to-hand and arm-to-arm contacts are followed by mouth-to-face and mouth-to-mouth ones. Mutual embracing occurs, both statically and during locomotion. Sudden spontaneous outbursts of running, chasing, jumping and dancing are commonly seen and juvenile play patterns may reappear.

Much of this pair-formation phase may take place in public, but when it passes over into the pre-copulatory phase, privacy is sought and the subsequent patterns of behaviour are performed in isolation from other members of the species as far as is possible. With the pre-copulatory stage there is a striking increase in the adoption of a horizontal posture. Body-to-body contacts are increased in both force and duration. Low-intensity side-by-side postures repeatedly give way to high-intensity face-to-face contacts. These positions may be maintained for many minutes and even for several hours, during which vocal and visual signals become gradually less important and tactile signals increasingly frequent. These involve small movements and varying pressures from all parts of the body, but in particular from the fingers, hands, lips and tongue. Clothing is partially or totally removed and skin-to-skin tactile stimulation is increased over as wide an area as possible.

Mouth-to-mouth contacts reach their highest frequency and their longest duration during this phase, the pressure exerted by the lips varying from extreme gentleness to extreme violence. During the higher-intensity responses the lips are parted and the tongue is inserted into the partner's mouth. Active movements of the tongue are then used to stimulate the sensitive skin of the mouth interior. The lips and tongue are also applied to many other areas of the partner's body, especially the ear-lobes, the neck and the genitals. The male pays particular attention to the breasts and nipples of the female, and the lip and tongue contact here becomes extended into more elaborate licking and sucking. Once contacted, the partner's genitals may also become the target for repeated actions of this kind. When this occurs, the male concentrates largely on the female's clitoris, the female on the male's penis, although other areas are also involved in both cases.

In addition to kissing, licking and sucking, the mouth is also applied to various regions of the partner's body in a biting action of varying intensities. Typically this involves no more than soft nibbling of the skin, or gentle nipping, but it can sometimes develop into forceful and even painful biting.

Interspersed between bouts of oral stimulation of the partner's body, and frequently accompanying it, there is a great deal of skin manipulation. The hands and fingers explore the whole body surface, concentrating especially on the face and, at higher intensities, on the buttocks and genital region. As in oral contacts, the male pays particular attention to the female's breasts and nipples. Wherever they move, the fingers repeatedly stroke and caress. From time to time they grasp with great force and the fingernails may be dug deeply into the flesh. The female may grasp the penis of the male, or stroke it rhythmically, simulating the movements of copulation, and the male stimulates the female genitals, especially the clitoris, in a similar way, again frequently with rhythmic movements.

In addition to these mouth, hand and general body contacts, there is also a tendency at high intensities of pre-copulatory activity to rub the genitals rhythmically against the partner's body. There is also a considerable amount of twining and inter-twining of the arms and legs, with occasional powerful muscle contractions, so that the body is thrown into a state of clinging tension, followed by relaxation.

These, then, are the sexual stimuli that are given to the partner during bouts of pre-copulatory activity, and which produce sufficient physiological sexual arousal for copulation to occur. Copulation starts with the insertion of the male's penis into the female's vagina. This is most commonly performed with the couple face-to-face, the male over the female, both in a horizontal position, with the female's legs apart. There are many variations of this position, as we shall be discussing later, but this is the simplest and most typical one. The male then begins a series of rhythmic pelvic thrusts. These can vary considerably in strength and speed, but in an uninhibited situation they are usually rather rapid and deeply penetrating. As copulation progresses there is a tendency to reduce the amount of oral and manual contact, or at least to reduce its subtlety and complexity. Nevertheless these now subsidiary forms of mutual stimulation do still continue to some extent throughout most copulatory sequences.

The copulatory phase is typically much briefer than the pre-copulatory phase. The male reaches the consummatory act of sperm ejaculation within a few minutes in most cases, unless deliberate delaying tactics are employed. Other female primates do not appear to experience a climax to their sexual sequences, but the naked ape is unusual in this respect. If the male continues to copulate for a longer period of time, the female also eventually reaches a consummatory moment, an explosive orgasmic experience, as violent and tension-releasing as the male's, and physiologically identical with it in every way except for the single obvious exception of sperm ejaculation. Some females may reach this point very quickly, others not at all, but on the average it is attained between ten and twenty minutes after the start of copulation.

It is strange that there is this discrepancy between the male and female as regards the time taken to reach sexual climax and relief from tension. This is a matter that will have to be discussed in detail later when the functional significance of the various sexual patterns are being considered. Suffice it to say at this point that the male can overcome the time factor and arouse the female to orgasm either by prolonging and heightening the pre-copulatory stimulation, so that she is already strongly aroused before penis insertion takes place, or he can employ self-inhibitory tactics during copulation to delay his own climax, or he can continue to copulate immediately after ejaculation and before he loses his erection, or he can rest briefly and then copulate for a second time. In the latter case, his reduced sex drive will automatically ensure that he takes much longer to reach his next climax and this will give the female sufficient time on this occasion to reach hers.

After both partners have experienced orgasm there normally follows a considerable period of exhaustion, relaxation, rest, and frequently sleep.

From the sexual stimuli we must now turn to the sexual responses. How does the body respond to all this intensive stimulation? In both sexes there are marked increases in pulse rate, blood pressure and respiration. These changes begin during pre-copulatory activities and rise to a peak at the copulatory climax. Pulse rates which, at normal level, stand at 70 to 80 per minute, rise to 90 to 100 during the earlier phases of sexual arousal, then climb to 130 during intense arousal and attain a peak of about 150 at orgasm. Blood pressure that starts at about 120 rises to 200 or even 250 at the sexual climax. Breathing becomes deeper and more rapid as arousal develops and then, as orgasm approaches, develops into prolonged gasping often accompanied by rhythmic moaning or grunting. At climax the face may be contorted, with mouth wide open and nostrils expanded, in a manner similar to that seen in an athlete in extremis, or someone fighting for air.

Another major change that occurs during sexual arousal is a dramatic shift in the distribution of blood, from the deeper regions to the surface areas of the body. This overall forcing of additional blood into the skin leads to a number of striking results. It produces not only a body that feels generally hotter to the touch—a sexual glow, or fire—but also certain specific changes in a number of specialized areas. At high intensities of arousal a characteristic sexual flush appears. It is most commonly seen in the female, where it usually begins in the region of skin over the stomach and upper abdomen, then spreads to the upper part of the breasts, then the upper chest, then the sides and middle region of the breasts and finally the undersides of the breasts. The face and neck may also be involved. In very intensely responding females it may also spread over the lower abdomen, the shoulders, the elbows, and, with orgasm, to the thighs, buttocks and back. In certain cases it may cover almost the whole body surface. It has been described as a measles-like rash and appears to be a visual sexual signal. It also occurs, but in fewer cases, in the male where, again, it starts in the region of the upper abdomer, spreads over the chest and then the neck and face. It occasionally also covers the shoulders, forearms and thighs. Once orgasm has been reached, the sex flush rapidly disappears, vanishing in reverse order to its sequence of appearance.

In addition to the sex flush and general vaso-dilation, there is also marked vaso-congestion of various distensible organs. This blood congestion is caused by the arteries pumping blood into these organs faster than the veins can carry it away. The condition can be maintained for considerable periods of time because the engorgement of the blood vessels in the organs itself helps to close off the veins that are attempting to carry the blood away. This occurs in the lips, nose, ear-lobes, nipples and genitals of both sexes and also in the breasts of the female. The lips become swollen, redder and more protuberant than at any other time. The soft parts of the nose become swollen and the nostrils expanded. The ear-lobes also become thickened and swollen. The nipples become enlarged and erect in both sexes, but more so in the female. (This is not due to vaso-congestion alone, but also to nipple muscle contraction.) Female nipple length increases by as much as one centimetre, and nipple diameter as much as a half a centimetre. The areola region of pigmented skin around the nipples also becomes tumescent and deeper in colour in the female, but not in the male. The female breast also shows a significant increase in size. By the time orgasm has been reached the breast of the average female will have increased by anything up to 25 per cent of its normal dimensions. It becomes firmer, more rounded and more protuberant.

The genitals of both sexes undergo considerable changes as arousal proceeds. The vaginal walls of the female experience massive vaso-congestion leading to rapid lubrication of the vaginal tube. In some cases this may occur within seconds of the beginning of pre-copulatory activity. There is also a lengthening and distension of the inner two-thirds of the vaginal tube, the overall length of the vagina increasing up to ten centimetres at the phase of high sexual excitement. As orgasm approaches, there is a swelling of the outer one-third of the vaginal tube, and during orgasm itself there is a two- to four-second muscle-spasm contraction of this region, followed by rhythmic contractions at intervals of 0·8 of a second. There are from three to fifteen of these rhythmic contractions in each orgasmic experience.

During arousal the external female genitals become considerably swollen. The outer labia open and swell, and may show size increases of up to two or three times the normal proportions. The inner labia also become distended to two or three times their normal diameter and they protrude through the protective curtain of the outer labia, adding as they do so an extra centimetre to the overall vaginal length. As arousal progresses there is a second striking change in the inner labia. Having already become vaso-congested and protuberant, they now change colour, turning bright red.

The clitoris (the female counterpart of the male penis) also becomes enlarged and more protuberant as sexual arousal begins, but as higher levels of excitement are reached, the labial swelling tends to mask this change and the clitoris is retracted under the labial hood. It cannot at this later stage be stimulated directly by the male's penis, but in its swollen and sensitive condition can still be affected indirectly by the rhythmic pressures applied to that region by the thrusting movements of the male.

The penis of the male undergoes a dramatic modification with sexual arousal. From a limp, flaccid condition it expands, stiffens and erects by means of intensive vaso-congestion. Its normal, average length of nine and a half centimetres is increased by seven to eight centimetres. The diameter is also considerably increased, giving the species the largest erect penis of any living primate.

At the moment of male sexual climax there are several powerful muscle contractions of the penis that expel the seminal fluid into the vaginal tube. The first of these contractions are the strongest ones and occur at intervals of 0·8 of a second—the same rate as the orgasmic vaginal contractions of the female.

During arousal the scrotal skin of the male becomes constricted and the mobility of the testes is reduced. They are elevated by a shortening of the spermatic cords (as, indeed, they are in states of cold, fear and anger) and are held tighter against the body. Vaso-congestion of the region results in a testicular size increase of up to fifty or even a hundred per cent.

These, then, are the principal ways in which the male and female bodies become modified by sexual activity. Once the climax has been reached, all the changes noted are rapidly reversed and the resting, post-sexual individual quickly returns to the normal quiescent physiological state. There is one final, post-orgasmic response that is worth mentioning. There may be a copious sweating by both male and female immediately following sexual climax and this may occur regardless of how much or how little physical effort has been put into the preceding sexual activities. However, although it is not related to total physical expenditure, it does bear a relationship to the intensity of the orgasm itself. The film of sweat develops on the back, the thighs and the upper chest. Sweat may run from the armpits. In intense cases, the whole of the trunk, from shoulders to thighs, may be involved. The palms of the hands and soles of the feet also perspire and, where the face has become mottled with the sexual flush, there may be sweating on the forehead and upper lip.

This brief summary of the sexual stimuli of our species and the responses given to them can now serve as a basis for discussing the significance of our sexual behaviour in relation to our ancestry and our general way of life, but first it is worth pointing out that the various stimuli and responses mentioned do not all occur with equal frequency. Some occur inevitably whenever a male and female come together for sexual activity, but others appear only in a proportion of the cases. Even so, they still occur with a sufficiently high frequency to be counted as 'species characteristics'. As regards the body responses, the sex flush is seen in 75 per cent of females and about 25 per cent of males. Nipple erection is universal for females, but only occurs in 60 per cent of males. Copious sweating after orgasm is a feature of 33 per cent of both males and females. Apart from these specific cases, most of the other body responses mentioned apply in all cases, although, of course, their actual intensity and duration will vary according to the circumstances.

Another point that requires clarification is the way in which these sexual activities are distributed throughout the individual's lifetime. During the first decade of life no true sexual activity can occur in either sex. A great deal of so-called 'sex-play' can be observed in young children, but until the female has begun to ovulate and the male to ejaculate, functional sexual patterns obviously cannot occur. Menstruation begins for some females at the age of ten and by the age of fourteen 80 per cent of young females are actively menstruating. All are doing so by the age of nineteen. The development of pubic hair, the broadening of the hips, and the swelling of the breasts accompanies this change and, in fact, slightly precedes it. General body growth proceeds at a slower rate and is not completed until the twenty-second year.

The first ejaculation in boys does not usually occur until they have reached eleven years, so that they are sexually slower starters than the girls. (The earliest recorded successful ejaculation is for a boy of eight, but this is most unusual.) By the age of twelve, 25 per cent of boys have experienced their first ejaculation and by fourteen 80 per cent have done so. (At this point, therefore, they have caught up with the girls.) The mean age for the first ejaculation is thirteen years and ten months. As with the girls, there are characteristic accompanying changes. Body hair begins to grow, especially in the pubic region and on the face. The typical sequence of appearance of this hairiness is: pubic, armpit, upper lip, cheeks, chin, and then, much more gradually, the chest and other parts of the body. Instead of a broadening of the hips, there is a widening of the shoulders. The voice becomes deeper. This last change also takes place in the girls but to a much smaller extent. In both sexes there is also an acceleration of the growth of the genital organs themselves.

It is interesting that, if one measures sexual responsiveness in term of frequency of orgasm, the male is much quicker to reach his peak of performance than the female. Although males begin their sexual maturation process a year or so behind the girls, they nevertheless attain their orgasmic peak while they are still in their teens, whereas the girls do not reach theirs until their mid-twenties or even thirties. In fact, the female of our species has to reach the age of twenty-nine before she can match the orgasm rate of the fifteen-year-old male. Only 23 per cent of fifteen-year-old females will have experienced orgasm at all, and this figure has only risen to 53 per cent by the age of twenty. By thirty-five it is 90 per cent.

The adult male achieves an average of about three orgasms a week, and over seven per cent experience daily or more than daily ejaculation. The frequency of orgasm for the average male is highest between the ages of fifteen and thirty, and then drops steadily from thirty to old age. The ability to achieve multiple ejaculation fades, and the angle at which the erect penis is carried also drops. Erection can be maintained for an average of nearly an hour in the late teens, but it has fallen to only seven minutes at the age of seventy. Nevertheless, 70 per cent of males are still sexually active at the age of seventy.

A similar picture of waning sexuality with increasing age

is found in the female. The more or less abrupt cessation of ovulation at around the age of fifty does not markedly reduce the degree of sexual responsiveness, when the population is taken as a whole. There are, however, great individual variations in its influence on sexual behaviour.

The vast majority of all the copulatory activity we have been discussing occurs when the partners are in a pair-bonded state. This may take the form of an officially recognized marriage, or an informal liaison of some sort. The high frequency of non-marital copulation that is known to take place should not be taken to imply a random promiscuity. In most cases it involves typical courtship and pair-formation behaviour, even if the resulting pair-bond is not particularly long-lasting. Approximately 90 per cent of the population becomes formally paired, but 50 per cent of females and 84 per cent of males will have experienced copulation before marriage. By the age of forty, 26 per cent of married females and 50 per cent of married males will have experienced extra-marital copulation. Official pair-bonds also break down completely in a number of cases and are abandoned (0·9 per cent in 1956 in America, for example). The pair-bonding mechanism in our species, although very powerful, is far from perfect.

Now that we have all these facts before us we can start to ask questions. How does the way we behave sexually help us to survive? Why do we behave in the way we do, rather than in some other way? We may be helped in these questions if we ask another one: How does our sexual behaviour compare with that of other living primates?

Straight away we can see that there is much more intense sexual activity in our own species than in any other primates, including our closest relations. For them, the lengthy courtship phase is missing. Hardly any of the monkeys and apes develop a prolonged pair-bond relationship. The pre-copu-latory patterns are brief and usually consist of no more than a few facial expressions and simple vocalizations. Copulation itself is also very brief. (In baboons, for instance, the time taken from mounting to ejaculation is no more than seven to eight seconds, with a total of no more than fifteen pelvic thrusts, often fewer.) The female does not appear to experience any kind of climax. If there is anything that could be called an orgasm it is a trivial response when compared with that of the female of our own species.

The period of sexual receptivity of the female monkey or ape is more restricted. It usually only lasts for about a week, or a little more, of their monthly cycle. Even this is an advance on the lower mammals, where it is limited more severely to the actual time of ovulation, but in our own species the pri-mate trend towards longer receptivity has been pushed to the very limit, so that the female is receptive at virtually all times. Once a female monkey or ape becomes pregnant, or is nursing a baby, she ceases to be sexually active. Again, our species has spread its sexual activities into these periods, so that there is only a brief time just before and just after parturition when mating is seriously limited.

Clearly, the naked ape is the sexiest primate alive. To find the reason for this we have to look back again at his origins. What happened? First, he had to hunt if he was to survive. Second, he had to have a better brain to make up for his poor hunting body. Third, he had to have a longer childhood to grow the bigger brain and to educate it. Fourth, the females had to stay put and mind the babies while the males went hunting. Fifth, the males had to co-operate with one another on the hunt. Sixth, they had to stand up straight and use weapons for the hunt to succeed. I am not implying that these changes happened in that order; on the contrary they undoubtedly all developed gradually at the same time, each modification helping the others along. I am simply enumerating the six basic, major changes that took place as the hunting ape evolved. Inherent in these changes there are, I believe, all the ingredients necessary to make up our present sexual complexity.

To begin with, the males had to be sure that their females were going to be faithful to them when they left them alone to go hunting. So the females had to develop a pairing tendency. Also, if the weaker males were going to be expected to co-operate on the hunt, they had to be given more sexual rights. The females would have to be more shared out, the sexual organization more democratic, less tyrannical. Each male, too, would need a strong pairing tendency. Furthermore, the males were now armed with deadly weapons and sexual rivalries would be much more dangerous: again, a good reason for each male being satisfied with one female. On top of that there were the much heavier parental demands being made by the slow-growing infants. Paternal behaviour would have to be developed and the parental duties shared between the mother and the father: another good reason for a strong pair-bond.

Given this situation as a starting point we can now see how other things grew from it. The naked ape had to develop the capacity for falling in love, for becoming sexually imprinted on a single partner, for evolving a pair-bond. Whichever way you put it, it comes to the same thing. How did he manage to do this? What were the factors that helped him in this trend? As a primate, he will already have had a tendency to form brief mateships lasting a few hours, or perhaps even a few days, but these now had to be intensified and extended. One thing that will have come to his aid is his own prolonged childhood. During the long, growing years he will have had the chance to develop a deep personal relationship with his parents, a relationship much more powerful and lasting than anything a young monkey could experience. The loss of this parental bond with maturation and independence would create a 'relationship void'—a gap that had to be filled. He would therefore already be primed for the development of a new, equally powerful bond to replace it.

Even if this was enough to intensify his need for forming a new pair-bond, there would still have to be additional assistance to maintain it. It would have to last long enough for the lengthy process of rearing a family. Having fallen in love, he would have to stay in love. By developing a prolonged and exciting courtship phase he could ensure the former, but something more would be needed after that. The simplest and most direct method of doing this was to make the shared activities of the pair more complicated and more rewarding. In other words, to make sex sexier.

How was this done? In every possible way, seems to be the answer. If we look back now at the behaviour of the present-day naked ape we can see the pattern taking shape. The increased receptivity of the female cannot be explained only in terms of increasing the birth-rate. It is true that by being prepared to copulate while still at the maternal phase of rearing a baby, the female does increase the birth-rate. With the very long dependency period, it would be a disaster if she did not. But this cannot explain why she is ready to receive the male and become sexually aroused throughout each of her monthly cycles. She only ovulates at one point during the cycle, so that mating at all other times can have no procreative function. The vast bulk of copulation in our species is obviously concerned, not with producing offspring, but with cementing the pair-bond by providing mutual rewards for the sexual partners. The repeated attainment of sexual consummation for a mated pair is clearly, then, not some kind of sophisticated, decadent outgrowth of modern civilization, but a deep-rooted, biologically based, and evolutionarily sound tendency of our species.

Even when she has stopped going through her monthly cycles — in other words, when she is pregnant — the female remains responsive to the male. This, too, is particularly important because, with a one-male-one-female system, it would be dangerous to frustrate the male for too long a period. It might endanger the pair-bond.

In addition to increasing the amount of time when sexual activities can take place, the activities themselves have been elaborated. The hunting life that gave us naked skins and more sensitive hands has given us much greater scope for sexually stimulating body-to-body contacts. During pre-copulatory behaviour these play a major role. Stroking, rubbing, pressing and caressing occur in abundance and far exceed anything found in other primate species. Also, special-ized organs such as the lips, ear-lobes, nipples, breasts and genitals are richly endowed with nerve-endings and have become highly sensitized to erotic tactile stimulation. The ear-lobes, indeed, appear to have been exclusively evolved to this end. Anatomists have often referred to them as meaning-less appendages, or 'useless, fatty excrescences'. In general parlance they are explained away as 'remnants' of the time when we had big ears. But if we look at other primate species, we find that they do not possess fleshy ear-lobes. It seems that, far from being a remnant, they are something new, and when we discover that, under the influence of sexual arousal, they become engorged with blood, swollen and hyper-sensitive, there can be little doubt that their evolution has been ex-clusively concerned with the production of yet another erogenous zone. (Surprisingly, the humble ear-lobe has been rather overlooked in this context, but it is worth noting that there are cases on record of both males and females actually reaching orgasm as a result of ear-lobe stimulation.) It is interesting to note that the protuberant, fleshy nose of our species is another unique and mysterious feature that the anatomists cannot explain. One has referred to it as a 'mere exuberant variation of no functional significance'. It is hard to believe that something so positive and distinctive in the way of primate appendages should have evolved without a function. When one reads that the side walls of the nose contain a spongy erectile tissue that leads to nasal enlargement and nostril expansion by vaso-congestion during sexual arousal, one begins to wonder.

As well as the improved tactile repertoire, there are some rather unique visual developments. Complex facial expressions play an important part here, although their evolution is concerned with improved communication in many other contexts as well. As a primate species we have the best developed and most complex facial musculature of the entire group. Indeed, we have the most subtle and complex facial expression system of all living animals. By making tiny movements of the flesh around the mouth, nose, eyes, eyebrows, and on the forehead, and by re-combining the movements in a wide variety of ways, we can convey a whole range of complex mood-changes. During sexual encounters, especially during the early courtship phase, these expressions are of paramount importance. (Their exact form will be discussed in another chapter.) Pupil dilation also occurs during sexual arousal and, although it is a small change, we may be more responsive to it than we realize. The eye-surface also glistens.

Like the ear-lobes and the protruding nose, the lips of our species are a unique feature, not found elsewhere in the primates. Of course, all primates have lips, but not turned inside-out like ours. A chimpanzee can protrude and turn back its lips in an exaggerated pout, exposing as it does so the mucous membrane that normally lies concealed inside the mouth. But the lips are only briefly held in this posture before the animal reverts to its normal 'thin-lipped' face. We, on the other hand, have permanently everted, rolled-back lips. To a chimpanzee we must appear to be in a permanent pout. If you ever have occasion to be embraced by a friendly chimpanzee, the kiss that it may then vigorously apply to your neck will leave you in no doubt about its ability to deliver a tactile signal with its lips. For the chimpanzee this is a greeting signal rather than a sexual one, but in our species it is used in both contexts, the kissing contact becoming particularly frequent and prolonged during the pre-copulatory phase. In connection with this development it was presumably more convenient to have the sensitive mucous surfaces permanently exposed, so that special muscle contractions around the mouth did not have to be maintained throughout the prolonged kissing contacts, but this is not the whole story. The exposed, mucous lips evolved a well-defined and characteristic shape. They did not grade off inconspicuously into the surrounding facial skin, but developed a fixed boundary line. In this way they also became important visual signalling devices. We have already seen that sexual arousal produces a swelling and reddening of the lips, and the clear demarcation of this area obviously assisted in the refinement of these signals, making subtle changes in lip condition more easily recognizable. Also, of course, even in their un-aroused condition they are redder than the rest of the face skin and, simply by their very existence, without indicating changes in physiological condition, they will act as advertising signals, drawing attention to the presence of a tactile sexual structure.

Puzzling over the significance of our unique mucous lips, anatomists have stated that their evolution 'is not as yet clearly understood' and have suggested that perhaps it has something to do with the increased amount of sucking that is required of the infant at the breast. But the young chimpanzee also does a great deal of very efficient sucking and its more muscular and prehensile lips would seem, if anything, to be better equipped for the job. Also, this cannot explain the evolution of a sharp margin between the lips and the surrounding face. Nor can it explain the striking differences in the lips of light- and dark-skinned populations. If, on the other hand, the lips are regarded as visual signalling devices, these differences are easier to understand. If climatic conditions demand a darker skin, then this will work against the visual signalling capacity of the lips by reducing their colour contrast. If they really are important as visual signals, then some kind of compensating development might be expected, and this is precisely what seems to have occurred, the negroid lips maintaining their conspicuousness by becoming larger and more protuberant. What they have lost in colour contrast, they have made up for in size and shape. Also, the margins of the negroid lips are more strikingly delineated. The 'lip-seams' of the paler races become exaggerated into more prominent ridges that are lighter in colour than the rest of the skin. Anatomically, these negroid characters do not appear to be primitive, but rather represent a positive advance in the specialization of the lip region.

There are a number of other obvious visual sexual signals. At puberty, as I have already mentioned, the arrival of a fully operative breeding condition is signalled by the development of conspicuous tufts of hair, especially in the region of the genitals and armpits and, in the male, on the face. In the female there is the rapid growth of the breasts. The body shape also changes, becoming broader at the shoulders in the male and at the pelvis in the female. These changes not only differentiate the sexually mature individual from the immature, but most of them also distinguish the mature male from the mature female. They not only act as signals revealing that the sexual system is now functional, but also indicate in each case whether it is masculine or feminine.

The enlarged female breasts are usually thought of primarily as maternal rather than sexual developments, but there seems to be little evidence for this. Other species of primates provide an abundant milk supply for their offspring and yet they fail to develop clearly defined hemispherical breast swellings. The female of our species is unique amongst primates in this respect. The evolution of protruding breasts of a characteristic shape appears to be yet another example of sexual signalling. This would be made possible and encouraged by the evolution of the naked skin. Swollen breast-patches in a shaggy-coated female would be far less conspicuous as signalling devices, but once the hair has vanished they would stand out clearly. In addition to their own conspicuous shape, they also serve to concentrate visual attention on to the nipples and to make the nipple erection that accompanies sexual arousal more conspicuous. The pigmented area of skin around the nipple, that deepens in colour during sexual arousal, also helps in the same way.

The nakedness of the skin also makes possible certain colour-change signals. These occur in limited areas in other animals where there are small naked patches, but have become more extensive in our own species. Blushing occurs with particularly high frequency during the earlier courtship stages of sexual behaviour and, at the later phases of more intense arousal, there is the characteristic mottling of the sex flush. (Again, this is a form of signalling that the darker-skinned races have had to sacrifice to climatic demands. We know that they still undergo these changes, however, because, although they are invisible as colour transformations, close examination reveals significant changes in skin texture.)

Before leaving this array of visual sexual signals we must consider a rather unusual aspect of their evolution. To do so, we must take a sidelong glance at some rather strange things that have been happening to the bodies of a number of our lowlier primate cousins, the monkeys. Recent German research has revealed that certain species have started to mimic themselves. The most dramatic examples of this are the mandrill and the gelada baboon. The male mandrill has a bright red penis with blue scrotal patches on either side of it. This colour arrangement is repeated on its face, its nose being bright red and its swollen, naked cheeks an intense blue. It is as if the animal's face is mimicking its genital region by giving the same set of visual signals. When the male mandrill approaches another animal, its genital display tends to be concealed by its body posture, but it can still apparently transmit the vital messages by using its phallic face. The female gelada indulges in a similar self-copying device. Around her genitals there is a bright red skin patch, bordered with white papillae. The lips of the vulva in the centre of this area are a deeper, richer red. This visual pattern is repeated on her chest region, where again there is a patch of naked red skin surrounded by the same kind of white papillae. In the centre of this chest patch the deep red nipples have come to lie so close together that they are strongly reminiscent of the lips of the vulva. (They are indeed so close to one another that the infant sucks from both teats at the same time.) Like the true genital patch, the chest patch varies in intensity of colour during the different stages of the monthly sexual cycle.

The inescapable conclusion is that the mandrill and the

gelada have brought their genital signals forward to a frontal position for some reason. We know too little about the life of mandrills in the wild to be able to speculate as to the reasons for this strange occurrence in this particular species, but we do know that wild geladas spend a great deal more of their time in an upright sitting posture than most other similar monkey species. If this is a more typical posture for them, then it follows that by having sexual signals on their chests they can more readily transmit these signals to other members of the group than if the markings only existed on their rear ends. Many species of primates have brightly coloured genitals, but these frontal mimics are rare.

Our own species has made a radical change in its typical body posture. Like geladas, we spend a great deal of time sitting up vertically. We also stand erect and face one another during social contacts. Could it be, then, that we, too, have indulged in something similar in the way of self-mimicry? Could our vertical posture have influenced our sexual signals? When considered in this way the answer certainly seems to be yes. The typical mating posture of all other primates involves the rear approach of the male to the female. She lifts her rear end and directs it towards the male. Her genital region is visually presented backwards to him. He sees it, moves towards her, and mounts her from behind. There is no frontal body contact during copulation, the male's genital region being pressed on to the female's rump region. In our own species the situation is very different. Not only is there pro-longed face-to-face pre-copulatory activity, but also copula-tion itself is primarily a frontal performance.

There has been some argument about this last point. It is a long-standing idea that the face-to-face mating position is the biologically natural one for our species, and that all others should be considered as sophisticated variations of it. Recent authorities have challenged this and have claimed that there is no such thing as a basic posture as far as we are concerned. Any body relationship, they feel, should be grist to our sexual mill, and as an inventive species it should be natural for us to experiment with any postures we like—the more the better, in fact, because this will increase the complexity of the sexual act, increase sexual novelty, and prevent sexual boredom between the members of a long-mated pair. Their argument is a perfectly valid one in the context within which they present it, but in trying to score their point they have gone too far. Their real objection was to the idea that any variations of the basic posture are 'sinful'. To counteract this idea, they stressed the value of these variations, and were quite right to do so, for the reasons given. Any improvement in sexual rewards for the members of a mated pair will obviously be important in strengthening the pair-bond. They are biologically sound for our species. But in fighting this battle the authorities concerned lost sight of the fact there is nevertheless one basic, natural mating posture for our species—the face-to-face posture. Virtually all the sexual signals and erogenous zones are on the front of the body—the facial expressions, the lips, the beard, the nipples, the areolar signals, the breasts of the female, the pubic hair, the genitals themselves, the major blushing areas, and the major sexual flush areas. It could be argued that many of these signals would operate perfectly well in the earlier stages, which could be face-to-face, but then, for the copulation itself, with both partners now fully aroused by frontal stimulation, the male could shift into a rear position for rear-entry copulation, or, for that matter, into any other unusual posture he cared to select. This is perfectly true, and possible as a novelty device, but it has certain disadvantages. To start with, the identity of the sexual partner is much more important to a pair-bonding species like ours. The frontal approach means that the in-coming sexual signals and rewards are kept tightly linked with the identity signals from the partner. Face-to-face sex is 'personalized sex'. In addition, the pre-copulatory tactile sensations from the frontally concentrated erogenous zones can be extended into the copulatory phase when the mating act is performed face-to-face. Many of these sensations would be lost by adopting other postures. Also, the frontal approach provides the maximum possibility for stimulation of the female's clitoris during the pelvic thrusting of the male. It is true that it will be passively stimulated by the pulling effect of the male's thrusts, regardless of his body position in relation to the female, but in a face-to-face mating there will in addition be the direct rhythmic pressure of the male's pubic region on to the clitoral area, and this will considerably heighten the stimulation. Finally, there is the basic anatomy of the female vaginal passage, the angle of which has swung forward to a marked degree, when compared with other species of primates. It has moved forward more than would be expected simply as a passive result of the process of becoming a vertical species. Undoubtedly, if it had been important for the female of our species to present her genitals to the male for rear mounting, natural selection would soon have favoured that trend and the females would by now have a more posteriorly directed vaginal tract.

So it seems plausible to consider that face-to-face copulation is basic to our species. There are, of course, a number of variations that do not eliminate the frontal element: male above, female above, side by side, squatting, standing, and so on, but the most efficient and commonly used one is with both partners horizontal, the male above the female. American investigators have estimated that in their culture 70 per cent of the population employ only this position. Even those who vary their postures still use the basic one for much of the time. Fewer than ten per cent experiment with rear-entry positions. In a massive, cross-cultural survey involving nearly two hundred different societies scattered all over the world, the conclusion was that copulation with the male entering the female from the rear does not occur as the usual practice in any of the communities studied.

If we can now accept this fact, we can return from this slight digression to the original question concerning sexual self-mimicry. If the female of our species was going to successfully shift the interest of the male round to the front, evolution would have to do something to make the frontal region more stimulating. At some point, back in our ancestry, we must have been using the rear approach. Supposing we had reached the stage where the female signalled sexually to the male from behind with a pair of fleshy, hemispherical buttocks (not, incidentally found elsewhere amongst the primates) and a pair of bright red genital lips, or labia. Supposing the male had evolved a powerful sexual responsiveness to these specific signals. Supposing that, at this point in evolution, the species became increasingly vertical and frontally orientated in its social contacts. Given this situation, one might very well expect to find some sort of frontal self-mimicry of the type seen in the gelada baboon. Can we, if we look at the frontal regions of the females of our species, see any structures that might possibly be mimics of the ancient genital display of hemispherical buttocks and red labia? The answer stands out as clearly as the female bosom itself. The protuberant, hemispherical breasts of the female must surely be copies of the fleshy buttocks, and the sharply defined red lips around the mouth must be copies of the red labia. (You may recall that, during intense sexual arousal, both the lips of the mouth and the genital labia become swollen and deeper in colour, so that they not only look alike, but also change in the same way in sexual excitement.) If the male of our species was already primed to respond sexually to these signals when they emanated posteriorly from the genital region, then he would have a built-in susceptibility to them if they could be reproduced in that form on the front of the female's body. And this, it would seem, is precisely what has happened, with the females carrying a duplicate set of buttocks and labia on their chests and mouths respectively. (The use of lipsticks and brassières immediately springs to mind, but these must be left until later, when we are dealing with the special sexual techniques of modern civilization.)

In addition to the all-important visual signals, there are certain odour stimuli that play a sexual role. Our sense of smell has been considerably reduced during evolution, but it is reasonably efficient and is more operative during sexual activities than we normally realize. We know that there are sex differences in body odours and it has been suggested that part of the process of pair-formation — falling in love — involves a kind of olfactory imprinting, a fixation on the specific individual odour of the partner's body. Connected with this is the intriguing discovery that at puberty there is a marked change in odour preferences. Before puberty there are strong preferences for sweet and fruity odours, but with the arrival of sexual maturity this response falls off and there is a dramatic shift in favour of flowery, oily and musky odours. This applies to both sexes, but the increase in musk responsiveness is stronger in males than females. It is claimed that as adults we can detect the presence of musk even when it is diluted down to one part in eight million parts of air, and it is significant that this substance plays a dominant role in the scent-signalling of many mammalian species, being produced in specialized scent-glands. Although we ourselves do not possess any large scent glands, we do have a large number of small ones—the apocrine glands. These are similar to ordinary sweat glands, but their secretions contain a higher proportion of solids. They occur on a number of parts of the body, but there are specially high concentrations of them in the regions of the armpits and the genitals. The localized hair-tufts that grow in these areas undoubtedly function as important scent-traps. It has been claimed that scent production in these areas is heightened during sexual arousal, but no detailed analysis of this phenomenon has yet been made. We do, however, know that there are 75 per cent more apocrine glands in the female of our species than in the male, and it is interesting to recall that in lower mammals during sexual encounters the male sniffs the female more than she sniffs him.

The location of our specialized odour-producing areas appears to be yet another adaptation to our frontal approach to sexual contact. There is nothing unusual about the genital centre, this we have in common with many other mammals, but the armpit concentration is a more unexpected feature. It appears to relate to the general tendency in our species to add new sexual stimulation centres to the front end of the body, in connection with the great increase in face-to-face sexual contacts. In this particular case it would result in the partner's nose being kept in close proximity with a major scent-producing area throughout much of the pre-copulatory and copulatory activity.

Up to this point we have been considering ways in which the appetitive sexual behaviour of our species has been improved and extended, so that contacts between the members of a mated pair have become increasingly rewarding, and their pair-bond therefore strengthened and maintained. But appetitive behaviour leads to a consummatory act and some improvements have been needed here too. Consider for a moment the old primate system. The adult males are sexually active all the time, except when they have just ejaculated. A consummatory orgasm is valuable for them because the relief from sexual tension that it brings damps down their sexual urges long enough for their sperm supplies to be replenished. The females, on the other hand, are sexually active only for a limited period centred around their ovulation time. During this period they are ready to receive the males at any time. The more copulations they experience, the greater the insurance that successful fertilization will be achieved. For them, there is no sexual satiation, no moment of copulatory climax that would pacify and tame their sexual urges. While they are on heat, there is no time to lose, they must keep going at all costs. If they experienced intense orgasms, they would then waste valuable potential mating time. At the end of a copulation, when the male ejaculates and dismounts, the female monkey shows little sign of emotional upheaval and usually wanders off as if nothing had happened.

With our own pair-bonding species the situation is entirely different. In the first place, as there is only a single male involved, there is no particular advantage in the female being sexually responsive at the point where he is sexually spent. So there is nothing working against the existence of a female orgasm. There are, however, two things working very much in its favour. One is the immense behavioural reward it brings to the act of sexual co-operation with the mated partner. Like all the other improvements in sexuality this will serve to strengthen the pair-bond and maintain the family unit. The other is that it considerably increases the chances of fertilization. It does this in a rather special way that applies only to our own peculiar species. Again, to understand this, we must look back at our primate relatives. When a female monkey has been inseminated by a male, she can wander away without any fear of losing the seminal fluid that now lies in the innermost part of her vaginal tract. She walks on all fours. The angle of her vaginal passage is still more or less horizontal. If a female of our own species were so unmoved by the experience of copulation that she too was likely to get up and wander off immediately afterwards, the situation would be different, for she walks bipedally and the angle of her vaginal passage during normal locomotion is almost vertical. Under the simple influence of gravity the seminal fluid would flow back down the vaginal tract and much of it would be lost. There is therefore a great advantage in any reaction that tends to keep the female horizontal when the male ejaculates and stops copulating. The violent response of female orgasm, leaving the female sexually satiated and exhausted, has precisely this effect. It is therefore doubly valuable.

The fact that the female orgasm in our species is unique amongst primates, combined with the fact that it is physiologically almost identical with the orgasmic pattern of the male, suggests that perhaps it is in an evolutionary sense a 'pseudo-male' response. In the make-up of both males and females there are latent properties belonging to the opposite sex. We know from comparative studies of other groups of animals that evolution can, if necessary, call up one of these latent qualities and bring it into the front line (in the 'wrong' sex, as it were). In this particular instance we know that the female of our species has developed a particular susceptibility to sexual stimulation of the clitoris. When we remember that this organ is the female homologue, or counterpart, of the male penis, this does seem to point to the fact that, in origin at any rate, the female orgasm is a 'borrowed' male pattern.

This may also explain why the male has the largest penis of any primate. It is not only extremely long when fully erect, but also very thick when compared with the penises of other species. (The chimpanzee's is a mere spike by comparison.) This broadening of the penis results in the female's external genitals being subjected to much more pulling and pushing during the performance of pelvic thrusts. With each inward thrust of the penis, the clitoral region is pulled downwards and then, with each withdrawal, it moves up again. Add to this the rhythmic pressure being exerted on to the clitoral region by the pubic region of the frontally copulating male, and you have a repeated massaging of the clitoris that—were she a male—would virtually be masturbatory.

So we can sum up by saying that with both appetitive and consummatory behaviour, everything possible has been done to increase the sexuality of the naked ape and to ensure the successful evolution of a pattern as basic as pair-formation, in a mammalian group where it is elsewhere virtually unknown. But the difficulties of introducing this new trend are not over yet. If we look at our naked ape couple, still successfully together and helping one another to rear the infants, all appears to be well. But the infants are growing now and soon they will have reached puberty, and then what? If the old primate patterns are left unmodified, the adult male will soon drive out the young males and mate with the young females. These will then become part of the family unit as additional breeding females along with their mother, and we shall be right back where we started. Also, if the young males are driven out into an inferior status on the edge of society, as in many primate species, then the co-operative nature of the all-male hunting group will suffer.

Clearly some additional modification to the breeding system is needed here, some kind of exogamy or out-breeding device. For the pair-bond system to survive, both the daughters and the sons will have to find mates of their own. This is not an unusual demand for pair-forming species and many examples of it can be found amongst the lower mammals, but the social nature of most primates makes it a more difficult proposition. In most pair-forming species the family splits up and spreads out when the young grow up. Because of its co-operative social behaviour the naked ape cannot afford to scatter in this way. The problem is therefore kept much more on the doorstep, but it is solved in basically the same way. As with all pair-bonded animals, the parents are possessive of one another. The mother 'owns' the father sexually and vice-versa. As soon as the offspring begin to develop their sexual signals at puberty, they become sexual rivals, the sons of the father and the daughters of the mother. There will be a tendency to drive them both out. The offspring will also begin to develop a need for a home-based 'territory' of their own. The urge to do this must obviously have been present in the parents for them to have set up a breeding home in the first place, and the pattern will simply be repeated. The parental home-base, dominated and 'owned' by the mother and father, will not have the right properties. Both the place itself and the individuals living in it will be heavily loaded with both primary and associative parental signals. The adolescent will automatically reject this and set off to establish a new breeding base. This is typical of young territorial carnivores, but not of young primates, and this is one more basic behavioural change that is going to be demanded of the naked ape.

It is perhaps unfortunate that this phenomenon of exogamy is so often referred to as indicating an 'incest taboo'. This immediately implies that it is a comparatively recent, culturally controlled restriction, but it must have developed biologically at a much earlier stage, or the typical breeding system of the species could never have emerged from its primate background.

Another related feature, and one that appears to be unique to our species, is the retention of the hymen or maidenhead in the female. In lower mammals it occurs as an embryonic stage in the development of the urogenital system, but as part of the naked ape's neoteny it is retained. Its persistence means that the first copulation in the life of the female will meet with some difficulty. When evolution has gone to such lengths to render her as sexually responsive as possible, it is at first sight strange that she should also be equipped with what amounts to an anti-copulatory device. But the situation is not as contradictory as it may appear. By making the first copulation attempt difficult and even painful, the hymen ensures that it will not be indulged in lightly. Clearly, during the adolescent phase, there is going to be a period of sexual experimentation, of 'playing the field' in search of a suitable partner. Young males at this time will have no good reason for stopping short of full copulation. If a pair-bond does not form they have not committed themselves in any way and can move on until they find a suitable mate. But if young females were to go so far without pair-formation, they might very well find themselves pregnant and heading straight towards a parental situation with no partner to accompany them. By putting a partial brake on this trend in the female, the hymen demands that she shall have already developed a deep emotional involvement before taking the final step, an involvement strong enough to take the initial physical discomfort in its stride.

A word must be added here on the question of monogamy and polygamy. The development of the pair-bond, which has occurred in the species as a whole, will naturally favour monogamy, but it does not absolutely demand it. If the violent hunting life results in adult males becoming scarcer than females, there will be a tendency for some of the surviving males to form pair-bonds with more than one female. This will then make it possible to increase the breeding rate without setting up dangerous tensions by creating 'spare' females. If the pair-formation process became so totally exclusive that it prevented this, it would be inefficient. This would not be an easy development, however, because of the possessiveness of the females concerned and the dangers of provoking serious sexual rivalries between them. Also working against it would be the basic economic pressures of maintaining the larger family group with all its offspring. A small degree of polygamy could exist, but it would be severely limited. It is interesting that although it still occurs in a number of minor cultures today, all the major societies (which account for the vast majority of the world population of the species) are monogamous. Even in those that permit polygamy, it is not usually practised by more than a small minority of the males concerned. It is intriguing to speculate as to whether its omission from almost all the larger cultures has, in fact, been a major factor in the attainment of their present successful status. We can, at any rate, sum up by saying that, whatever obscure, backward tribal units are doing today, the mainstream of our species expresses its pair-bonding character in its most extreme form, namely long-term monogamous matings.

This, then, is the naked ape in all its erotic complexity: a highly sexed, pair-forming species with many unique features; a complicated blend of primate ancestry with extensive carnivore modifications. Now, to this we must add the third and final ingredient: modern civilization. The enlarged brain that accompanied the transformation of the simple forest-dweller into a co-operative hunter began to busy itself with technological improvements. The simple tribal dwelling places became great towns and cities. The axe age blossomed into the space age. But what effect did the acquisition of all this gloss and glitter have on the sexual system of the species? Very little, seems to be the answer. It has all been too quick, too sudden, for any fundamental biological advances to occur. Superficially they *seem* to have occurred, it is true, but this is largely make-believe. Behind the façade of modern city life there is the same old naked ape. Only the names have been changed: for 'hunting' read 'working', for 'hunting grounds' read 'place of business', for 'home base' read 'house', for 'pair-bond' read 'marriage', for 'mate' read 'wife', and so on. The American studies of contemporary sexual patterns, referred to earlier, have revealed that the physiological and anatomical equipment of the species is still being put to full use. The evidence of prehistoric remnants combined with comparative data from living carnivores and other living primates has given us a picture of how the naked ape must have used this sexual equipment in the distant past and how he must have organized his sex life. The contemporary evidence appears to give much the same basic picture, once one has cleaned away the dark varnish of public moralizing. As I said at the beginning of the chapter, it is the biological nature of the beast that has moulded the social structure of civilization, rather than the other way around.

Yet, although the basic sexual system has been retained in a fairly primitive form (there has been no communalization of sex to match the enlarged communities), many minor controls and restrictions have been introduced. These have become necessary because of the elaborate set of anatomical and physiological sexual signals and the heightened sexual responsiveness we have acquired during our evolution. But these were designed for use in a small, closely knit tribal unit, not in a vast metropolis. In the big city we are constantly intermixing with hundreds of stimulating (and stimulatable) strangers. This is something new, and it has to be dealt with.

In fact, the introduction of cultural restrictions must have begun much earlier, before there were strangers. Even in the simple tribal units it must have been necessary for members of a mated pair to curtail their sexual signalling in some way when they were moving about in public. If sexuality had to be heightened to keep the pair together, then steps must have been taken to damp it down when the pair were apart, to avoid the over-stimulation of third parties. In other pair-forming but communal species this is done largely by aggressive gestures, but in a co-operative species like ours, less belligerent methods would be favoured. This is where our enlarged brain can come to the rescue. Communication by speech obviously plays a vital role here ('My husband wouldn't like it'), as it does in so many facets of social contact, but more immediate measures are also needed.

The most obvious example is the hallowed and proverbial fig-leaf. Because of his vertical posture it is impossible for a naked ape to approach another member of his species without performing a genital display. Other primates, advancing on all fours, do not have this problem. If they wish to display their genitals they have to assume a special posture. We are faced with it, hour in and hour out, whatever we are doing. It follows that the covering of the genital region with some simple kind of garment must have been an early cultural development. The use of clothing as a protection against the cold no doubt developed from this as the species spread its range into less friendly climates, but that stage probably came much later.

With varying cultural conditions, the spread of the anti-sexual garments has varied, sometimes extending to other secondary sexual signals (breast coverings, lip-veils), sometimes not. In certain extreme cases the genitals of the females are not only concealed but also made completely inaccessible. The most famous example of this is the chastity belt, which covered the genital organs and anus with a metal band perforated in the appropriate places to permit the passage of body excretions. Other similar practices have included the sewing up of the genitals of young girls before marriage, or the securing of the labia with metal clasps or rings. In more recent times a case has been recorded of a male boring holes in his mate's labia and then padlocking her genitals after each copulation. Such extreme precautions as this are, of course, very rare, but the less drastic course of simply hiding the genitals behind a concealing garment is now almost universal.

Another important development was the introduction of privacy for the sexual acts themselves. The genitals not only became private parts, they also had to be privately used parts. Today this has resulted in the growth of a strong association between mating and sleeping activities. Sleeping with someone has become synonymous with copulating with them: so, the vast bulk of copulatory activity, instead of being spread out through the day, has now become limited to one particular time – the late evening.

Body-to-body contacts have, as we have seen, become such an important part of sexual behaviour that these too have to be damped down during the ordinary daily routine. A ban has to be placed on physical contact with strangers in our busy, crowded communities. Any accidental brushing against a stranger's body is immediately followed by an apology, the intensity of this apology being proportional to the degree of sexuality of the part of the body touched. Speeded-up film of a crowd moving through a street, or milling around in a large building, reveals clearly the incredible intricacy of these non-stop 'bodily-contact avoidance' manœuvres.

This contact restriction with strangers normally breaks down only under conditions of extreme crowding, or in special circumstances in relation to particular categories of individuals (hairdressers, tailors and doctors, for instance) who are socially 'licensed to touch'. Contact with close friends and relatives is less inhibited. Their social roles are already clearly established as non-sexual and there is less danger. Even so, greeting ceremonies have become highly stylized. The hand-shake has become a rigidly fixed pattern. The greeting kiss has developed its own ritualized form (mutual mouth-to-cheek touching) that sets it apart from mouth-to-mouth sexual kissing.

Body postures have become de-sexualized in certain ways. The female sexual invitation posture of legs apart is strongly avoided. When sitting, the legs are kept tightly together, or crossed one over the other.

If the mouth is forced to adopt a posture that is reminiscent in some way of a sexual response, it is often hidden by the hand. Giggling and certain kinds of laughing and grimacing are characteristic of the courtship phase, and when these occur in social contexts, the hand can frequently be seen to shoot up and cover the mouth region.

Males in many cultures remove certain of their secondary sexual characters by shaving off their beards and/or moustaches. Females depilate their armpits. As an important scent-trap, the armpit hair-tuft has to be eliminated if normal dressing habits leave that region exposed. Pubic hair is always so carefully concealed by clothing that it does not usually warrant this treatment, but it is interesting that this area is also frequently shaved by artists' models, whose nudity is non-sexual.

In addition, a great deal of general body de-scenting is practised. The body is washed and bathed frequently—far more than is required simply by the demands of medical care and hygiene. Body odours are socially suppressed and commercial chemical deodorants are sold in large numbers.

Most of these controls are maintained by the simple, unanswerable strategy of referring to the phenomena they restrict as 'not nice', 'not done', or 'not polite'. The true anti-sexual nature of the restrictions is seldom mentioned or even considered. More overt controls are also applied, however, in the form of artificial moral codes, or sexual laws. These vary considerably from culture to culture, but in all cases the major concern is the same – to prevent sexual arousal of strangers and to curtail sexual inter-action outside the pair-bond. As an aid to this process, which is recognized to be a difficult one even by the most puritanical groups, various sublimatory techniques are employed. Schoolboy sports, for example, and other vigorous physical activities are sometimes encouraged in the vain hope that this will reduce the sexual urges. Careful examination of this concept and its application reveals that by and large it is a dismal failure. Athletes are neither more nor less sexually active than other groups. What they lose from physical exhaustion, they gain in physical fitness. The only behavioural method that seems to be of assistance is the age-old system of punishment and reward – punishment for sexual indulgence and reward for sexual restraint. But this, of course, produces suppression rather than reduction of drive.

It is quite clear that our unnaturally enlarged communities will call for some steps of this kind to prevent the intensified social exposure from leading to dangerously increased sexual activities outside the pair-bond. But the naked ape's evolution as a highly sexed primate can take only so much of this treatment. Its biological nature keeps on rebelling. As fast as artificial controls are applied in one way, counter-acting improvements are made in another. This often leads to ridiculously contradictory situations.

The female covers her breasts, and then proceeds to redefine their shape with a brassière. This sexual signalling device may be padded or inflatable, so that it not only reinstates the concealed shape, but also enlarges it, imitating in this way the breast-swelling that occurs during sexual arousal. In some cases, females with sagging breasts even go to the length of cosmetic surgery, subjecting themselves to sub-cutaneous wax injections to produce similar effects on a more permanent basis.

Sexual padding has also been added to certain other parts of the body: one only has to think of male codpieces and padded shoulders, and female buttock-enlarging bustles. In certain cultures today it is possible for skinny females to purchase padded buttock-brassières, or 'bottom-falsies'. The wearing of high-heeled shoes, by distorting the normal walking posture, increases the amount of swaying in the buttock region during locomotion.

Female hip-padding has also been employed at various times and, by the use of tight belts, both the hip and breast curves can be exaggerated. Because of this, small female waists have been strongly favoured and tight corseting of this region has been widely practised. As a trend this reached its peak with the 'wasp waists' of half a century ago, at which time some females even went to the extreme of having the lower ribs removed surgically to increase the effect.

The widespread use of lipstick, rouge and perfume to heighten sexual lip signals, flushing signals, and body-scent signals respectively, provide further contradictions. The female who so assiduously washes off her own biological scent then proceeds to replace it with commercial 'sexy' perfumes which, in reality, are no more than diluted forms of the products of the scent-glands of other, totally unrelated mammalian species.

Reading through all these various sexual restrictions and the artificial counter-attractions, one cannot help feeling that it would be much easier simply to go back to square one. Why refrigerate a room and then light a fire in it? As I explained before, the reason for the restrictions is straight-forward enough: it is a matter of preventing random sexual stimulation which would interfere with the pair-bonds. But why not a total restriction in public? Why not limit the sexual displays, both biological and artificial, to the moments of privacy between the members of the mated pair? Part of the answer to this is our very high level of sexuality, which demands constant expression and outlet. It was developed to keep the pair together, but now, in the stimulating atmosphere of a complex society, it is constantly being triggered off in non-pair-bond situations. But this is only part of the answer. Sex is also being used as a status device—a well-known manœuvre in other primate species. If a female monkey wants to approach an aggressive male in a non-sexual context, she may display sexually to him, not because she wants to copu-late, but because by so doing she will arouse his sexual urges sufficiently to suppress his aggression. Such behaviour patterns are referred to as re-motivating activities. The female uses sexual stimulation to re-motivate the male and thereby gain a non-sexual advantage. Similar devices are used by our own species. Much of the artificial sexual signalling is being employed in this way. By making themselves attractive to members of the opposite sex, individuals can effectively reduce antagonistic feelings in other members of the social group.

There are dangers in this strategy, of course, for a pair-bonding species. The stimulation must not go too far. By conforming to the basic sexual restrictions that the culture has developed, it is possible to give clear signals that 'I am not available for copulation', and yet, at the same time, to give other signals which say that 'I am nevertheless very sexy'. The latter signals will do the job of reducing antagonism, while the former ones will prevent things from getting out of hand. In this way one can have one's cake and eat it.

This should work out very neatly, but unfortunately there are other influences at work. The pair-bonding mechanism is not perfect. It has had to be grafted on to the earlier primate system, and this still shows through. If anything goes wrong with the pair-bond situation, then the old primate urges flare up again. Add to this the fact that another of the naked ape's great evolutionary developments has been the extension of childhood curiosity into the adult phase, and the situation can obviously become dangerous.

The system was obviously designed to work in a situation where the female is producing a large family of overlapping children and the male is off hunting with other males. Although fundamentally this has persisted, two things have changed. There is a tendency to limit artificially the number of off-spring. This means that the mated female will not be at full parental pressure and will be more sexually available during her mate's absence. There is also a tendency for many females to join the hunting group. Hunting, of course, has now been replaced by 'working' and the males who set off on their daily working trips are liable to find themselves in hetero-sexual groups instead of the old all-male parties. This means that the pair-bond has a lot to put up with on both sides. All too often it collapses under the strain. (The American figures, you will recall, indicated that 26 per cent of married females and 50 per cent of married males have experienced extra-marital copulation by the age of forty.) Frequently, though, the original pair-bond is strong enough to maintain itself during these outside activities, or to re-assert itself when they have passed. In only a small percentage of cases is there a complete and final break-down.

To leave the matter there would overstate the case for the pair-bond, however. It may be able to survive sexual curiosity in most cases, but it is not strong enough to stamp it out. Although the powerful sexual imprinting keeps the mated pair together, it does not eliminate their interest in outside sexual activities. If outside matings conflict too strongly with the pair-bond, then some less harmful substitute for them has to be found. The solution has been voyeurism, using the term in its broadest sense, and this is employed on an enormous scale. In the strict sense voyeurism means obtaining sexual excitement from watching other individuals copulating, but it can logically be broadened out to include any non-partici-patory interest in any sexual activity. Almost the entire population indulges in this. They watch it, they read about it, they listen to it. The vast bulk of all television, radio, cinema, theatre and fiction-writing is concerned with satisfying this demand. Magazines, newspapers and general conversation also make a large contribution. It has become a major industry. And never once throughout all this does the sexual observer actually *do* anything. Everything is performed by proxy. So urgent is the demand, that we have had to invent a special category of performers—actors and actresses—who pretend to go through sexual sequences for us, so that we can watch them at it. They court and marry, and then live again in new roles, to court and marry another day. In this way the voyeur supplies are tremendously increased.

If one looked at a wide range of animal species, one would be forced to the conclusion that this voyeurist activity of ours is biologically abnormal. But it is comparatively harmless and may actually help our species, because it satisfies to some extent the persistent demands of our sexual curiosity without involving the individuals concerned in new potential mateship relationships that could threaten the pair-bond.

Prostitution operates in much the same way. Here, of course, there is involvement, but in the typical situation it is ruthlessly restricted to the copulatory phase. The earlier court-ship phase and even the pre-copulatory activities are kept to an absolute minimum. These are the stages where pair-formation begins to operate and they are duly suppressed. If a mated male indulges his urge for sexual novelty by copulat-ing with a prostitute he is, of course, liable to damage his pair-bond, but less so than if he becomes involved in a romantic, but non-copulatory, love affair.

Another form of sexual activity that requires examination is the development of a homosexual fixation. The primary function of sexual behaviour is to reproduce the species and this is something that the formation of homosexual pairs patently fails to do. It is important to make a subtle distinction here. There is nothing biologically unusual about a homo-sexual act of pseudo-copulation. Many species indulge in this, under a variety of circumstances. But the formation of a homosexual pair-bond is reproductively unsound, since it cannot lead to the production of offspring and wastes potential breeding adults. To understand how this can happen it will help to look at other species.

I have already explained how a female may use sexual signals to re-motivate an aggressive male. By arousing him sexually she suppresses his antagonism and avoids being attacked. A subordinate male may use a similar device. Young male monkeys frequently adopt female sexual in-vitation postures and are then mounted by dominant males that would otherwise have attacked them. Dominant females may also mount subordinate females in the same way. This utilization of sexual patterns in non-sexual situations has become a common feature of the primate social scene and has proved extremely valuable in helping to maintain group harmony and organization. Because these other species of primates do not undergo a process of intense pair-bond formation, it does not lead to difficulties in the shape of long-term homosexual pairings. It simply solves immediate dominance problems, but does not have long-term sexual relationship consequences.

Homosexual behaviour is also seen in situations where the ideal sexual object (a member of the opposite sex) is unavailable. This applies in many groups of animals: a member of the same sex is used as a substitute object — 'the next best thing' for sexual activity. In total isolation animals are often driven to more extreme measures and will attempt to copulate with inanimate objects, or will masturbate. In captivity, for example, certain carnivores have been known to copulate with their food containers. Monkeys frequently develop masturbatory patterns and this has even been recorded in the case of lions. Also, animals housed with the wrong species may attempt to mate with them. But these activities typically disappear when the biologically correct stimulus — a member of the opposite sex — appears on the scene.

Similar situations occur with high frequency in our own species and the response is much the same. If either males or females cannot for some reason obtain sexual access to their opposite numbers, they will find sexual outlets in other ways. They may use other members of their own sex, or they may even use members of other species, or they may masturbate. The detailed American studies of sexual behaviour revealed that in that culture 13 per cent of females and 37 per cent of males have indulged in homosexual contacts to the point of orgasm by the age of 45. Sexual contacts with other animal species are much rarer (because, of course, they provide far fewer of the appropriate sexual stimuli) and have been recorded in only 3·6 per cent of females and 8 per cent of males. Masturbation, although it does not provide 'partner stimuli', is nevertheless so much easier to initiate that it occurs with a much higher frequency. It is estimated that 58 per cent of females and 92 per cent of males masturbate at some time in their lives.

If all these reproductively wasteful activities can take place without reducing the long-term breeding potential of the individuals concerned, then they are harmless. In fact, they can be biologically advantageous, because they can help to prevent sexual frustration which can lead to social disharmony in various ways. But the moment they give rise to sexual fixations they create a problem. In our species there is, as we have seen, a strong tendency to 'fall in love' — to develop a powerful bond with the object of our sexual attentions. This sexual imprinting process produces the all-important long-term mateship so vital to the prolonged parental demands. The imprinting is going to start operating as soon as serious sexual contacts are made, and the consequences are obvious. The earliest objects towards which we direct our sexual attentions are liable to become *the* objects. Imprinting is an associative process. Certain key stimuli that are present at the moment of sexual reward become intimately linked with the reward, and in no time at all it is impossible for sexual behaviour to occur without the presence of these vital stimuli. If we are driven by social pressures to experience our earliest sexual rewards in homosexual or masturbatory contexts, then certain elements present in these contexts are likely to assume powerful sexual significance of a lasting kind. (The more unusual forms of fetishism also originate in this way.)

One might expect these facts to lead to more trouble than

actually occurs, but two things help to prevent this in most cases. Firstly, we are well equipped with a set of instinctive responses to the characteristic sexual signals of the opposite sex, so that we are unlikely to experience a powerful courtship reaction to any object lacking these signals. Secondly, our earliest sexual experiments are of a very tentative nature. We start by falling in and out of love very frequently and very easily. It is as if the process of full imprinting lags behind the other sexual developments. During this 'searching' phase we typically develop a large number of minor 'imprints', each one being counteracted by the next, until eventually we arrive at a point where we are susceptible to a major imprinting. Usually by this time we have been sufficiently exposed to a variety of sexual stimuli to have latched on to the appropriate biological ones, and mating then proceeds as a normal hetero-sexual process.

It will perhaps be easier to understand this if we compare it with the situation that has evolved in certain other species. Pair-forming colonial birds, for example, migrate to the breeding grounds where the nest sites will be established. Young and previously unmated birds, flying in as adults for the first time, must, like all the older birds, establish territories and form breeding pairs. This is done without much delay, soon after arrival. The young birds will select mates on the basis of their sexual signals. Their response to these signals will be inborn. Having courted a mate they will then limit their sexual advances to that particular individual. This is achieved by a process of sexual imprinting. As the pair-forming court-ship proceeds, the instinctive sexual clues (which all members of each sex of each species will have in common) have to become linked with certain unique individual recognition characters. Only in this way can the imprinting process narrow down the sexual responsiveness of each bird to its mate. All this has to be done quickly, because the breeding season is limited. If, at the start of this stage, all members of one sex were experimentally removed from the colony, a large number of homosexual pair-bonds might become established, as the birds desperately tried to find the nearest thing to a correct mate that was available.

In our own species the process is much slower. We do not have to work against the deadline of a brief breeding season. This gives us time to scout around and 'play the field'. Even if we are thrown into a sexually segregated environment for considerable periods during adolescence, we do not all automatically and permanently develop homosexual pair-bonds. If we were like the colonial nesting birds, then no young male could emerge from an all-male boarding school (or other similar unisexual organization) with the slightest hope of ever making a heterosexual pair-bond. As it is, the process is not too damaging. The imprinting canvas is only lightly sketched in in most cases and can easily be erased by later, more powerful impressions.

In a minority of cases, however, the damage is more permanent. Powerful associative features will have become firmly linked with sexual expression and will always be required in later bond-forming situations. The inferiority of the basic sexual signals given by a partner of the same sex will not be sufficient to outweigh the positive imprinting associations. It is a fair question to ask why a society should expose itself to such dangers. The answer seems to be that it is caused by a need to prolong the educational phase as much as possible to cope with the enormously elaborated and complicated technological demands of the culture. If young males and females established family units as soon as they were biologically equipped to do so, a great deal of training potential would be wasted. Strong pressures are therefore put upon them to prevent this. Unfortunately, no amount of cultural restriction is going to prevent the development of the sexual system, and if it cannot take the usual route it will find some other.

There is another separate but important factor that can influence homosexual trends. If, in the parental situation, the offspring are exposed to an unduly masculine and dominant mother, or an unduly weak and effeminate father, then this will give rise to considerable confusion. Behavioural characters will point one way, anatomical ones the other. If, when they become sexually mature, the sons seek mates with the behavioural (rather than the anatomical) qualities of the mother, they are liable to take male mates rather than females. For the daughters there is a similar risk, in reverse. The trouble with sexual problems of this sort is that the prolonged period of infant dependency creates such an enormous overlap between the generations, that disturbances are carried over, time after time. The effeminate father mentioned above was probably previously exposed to sexual abnormalities in the relationship between his own parents, and so on. Problems of this kind reverberate down the generations for a long time before they peter out, or before they become so acute that they solve themselves by preventing breeding altogether.

As a zoologist I cannot discuss sexual 'peculiarities' in the usual moralistic way. I can only apply anything like a biological morality in terms of population success and failure. If certain sexual patterns interfere with reproductive success, then they can genuinely be referred to as biologically unsound. Such groups as monks, nuns, long-term spinsters and bachelors and permanent homosexuals are all, in a reproductive sense, aberrant. Society has bred them, but they have failed to return the compliment. Equally, however, it should be realized that an active homosexual is no more reproductively aberrant than a monk. It must also be said that no sexual practice, no matter how disgusting and obscene it may appear to the members of a particular culture, can be criticized biologically providing it does not hinder general reproductive success. If the most bizarre elaboration of sexual performance helps to ensure either that fertilization will occur between members of a mated pair, or that the pair-bond will be strengthened, then reproductively it has done its job and is biologically just as acceptable as the most 'proper' and approved-of sexual customs.

Having said all this, I must now point out that there is an important exception to the rule. The biological morality that I have outlined above ceases to apply under conditions of population over-crowding. When this occurs the rules become reversed. We know from studies of other species in experimentally over-crowded conditions that there comes a moment when the increasing population density reaches such a pitch that it destroys the whole social structure. The animals develop diseases, they kill their young, they fight viciously and they mutilate themselves. No behaviour sequence can run through properly. Everything is fragmented. Eventually there are so many deaths that the population is cut back to a lower density and can start to breed again, but not before there has been a catastrophic upheaval. If, in such a situation, some controlled anti-reproductive device could have been introduced into the population when the first signs of over-crowding were apparent, then the chaos could have been averted. Under such conditions (serious over-crowding with no signs of any easing up in the immediate future), anti-reproductive sexual patterns must obviously be considered in a new light.

Our own species is rapidly heading towards just such a situation. We have arrived at a point where we can no longer be complacent. The solution is obvious, namely to reduce the breeding rate without interfering with the existing social structure; to prevent an increase in quantity without preventing an increase in quality. Contraceptive techniques are obviously required, but they must not be allowed to disrupt the basic family unit. Actually there should be little danger of this. Fear has been expressed that the widespread use of perfected contraceptives will lead to random promiscuity, but this is most unlikely — the powerful pair-formation tendency of the species will see to that. There may be some trouble if many mated pairs employ contraception to the point where no offspring are produced. Such couples will put heavy demands on their pair-bonds, which may break under the strain. These individuals will then constitute a greater threat to other pairs that are attempting to rear families. But extreme breeding reductions of this kind are not necessary. If every family produced two children, the parents would simply reproduce their own number and there would be no increase. Allowing for accidents and premature deaths, the average figure could be slightly higher than this without leading to further population increase and eventual species catastrophe.

The trouble is that, as a sexual phenomenon, mechanical and chemical contraception is something basically new and it will take some time before we know exactly what sort of repercussions it will have on the fundamental sexual structure of society after a large number of generations have experienced it and new traditions have gradually developed out of old ones. It may cause indirect, unforeseen distortions or disruptions of the socio-sexual system. Only time will tell. But whatever happens the alternative, if breeding limitation is not applied, is far worse.

Bearing in mind this over-crowding problem, it could be argued that the need to reduce drastically the reproduction rate now removes any biological criticism of the non-breeding categories such as the monks and nuns, the long-term spinsters and bachelors, and the permanent homosexuals. Purely on a reproductive basis this is true, but it leaves out of account the other social problems that, in certain cases, they may have to face, set aside in their special minority roles. Nevertheless, providing they are well adjusted and valuable members of society outside the reproductive sphere, they must now be considered as valuable non-contributors to the population explosion.

Looking back now on the whole sexual scene we can see that our species has remained much more loyal to its basic biological urges than we might at first imagine. Its primate sexual system with carnivore modifications has survived all the fantastic technological advances remarkably well. If one took a group of twenty suburban families and placed them in a primitive sub-tropical environment where the males had to go off hunting for food, the sexual structure of this new tribe would require little or no modification. In fact, what has happened in every large town or city is that the individuals it contains have specialized in their hunting (working) techniques, but have retained their socio-sexual system in more or less its original form. Science-fiction conceptions of baby-farms, communalized sexual activities, selective sterilization, and state-controlled division of labour in reproductive duties, have not materialized. The space ape still carries a picture of his wife and children with him in his wallet as he speeds towards the moon. Only in the field of general breeding limitation are we now coming face to face with the first major assault on our age-old sexual system by the forces of modern civilization. Thanks to medical science, surgery and hygiene, we have reached an incredible peak of breeding success. We have practised death control and now we must balance it with birth control. It looks very much as though, during the next century or so, we are going to have to change our sexual ways at last. But if we do, it will not be because they failed, but because they succeeded too well.

Chapter Three

*

REARING

The burden of parental care is heavier for the naked ape than for any other living species. Parental duties may be performed as *in*tensively elsewhere, but never so *ex*tensively. Before we consider the significance of this trend, we must assemble the basic facts.

Once the female has been fertilized and the embryo has started to grow in her uterus, she undergoes a number of changes. Her monthly menstrual flow ceases. She experiences early-morning nausea. Her blood pressure is lower. She may become slightly anaemic. As time passes, her breasts become swollen and tender. Her appetite increases. Typically she becomes more placid.

After a gestation period of approximately 266 days her uterus begins to contract powerfully and rhythmically. The amniotic membrane surrounding the foetus is ruptured and the fluid in which the baby has been floating escapes. Further violent contractions expel the infant from the womb, forcing it through the vaginal passage and into the outside world. Renewed contractions then dislodge and eject the placenta. The cord connecting the baby to the placenta is then severed. In other primates this breaking of the cord is achieved by the mother biting through it, and this was no doubt the method employed by our own ancestors, but today it is neatly tied and snipped through with a pair of scissors. The stump still attached to the infant's belly dries up and drops off a few days after birth.

It is a universal practice today for the female to be accompanied and aided by other adults while she is giving birth. This is probably an extremely ancient procedure. The demands of vertical locomotion have not been kind to the female of our species: the penalty for this progressive step is a sentence of several hours' hard labour. It seems likely that co-operation from other individuals was needed right back at the stages where the hunting ape was evolving from its forest-dwelling ancestors. Luckily the co-operative nature of the species was growing alongside this hunting development, so that the cause of the trouble could also provide its cure. Normally, the chimpanzee mother not only bites through the cord, she also devours all or part of the placenta, licks up the fluids, washes and cleans her newly delivered infant, and holds it protectively to her body. In our own species the exhausted mother relies on companions to perform all these activities (or modern equivalents of them).

After the birth is over it may take a day or two for the mother's milk-flow to get started, but once this has happened she then feeds the baby regularly in this way for a period of up to two years. The average suckling period is shorter than this, however, and modern practice has tended to reduce it to six to nine months. During this time the menstrual cycle of the female is normally suppressed and the menstrual flow usually reappears only when she stops breast-feeding and starts to wean the baby. If infants are weaned unusually early, or if they are bottle-fed, this delay does not, of course, occur, and the female can start breeding again more quickly. If, on the other hand, she follows the more primitive system and feeds the infant for a full two-year period, she is liable to produce offspring only about once in three years. (Suckling is sometimes deliberately prolonged in this way as a contraceptive technique.) With a reproductive life-span of approximately thirty years, this puts her natural productivity capacity at about ten offspring. With bottle-feeding or rapidly curtailed breast-feeding, the figure could theoretically rise to thirty.

The act of suckling is more of a problem for females of our species than for other primates. The infant is so helpless that the mother has to take a much more active part in the process, holding the baby to the breast and guiding its actions. Some mothers have difficulty in persuading their offspring to suck efficiently. The usual cause of this trouble is that the nipple is not protruding far enough into the baby's mouth. It is not enough for the infant's lips to close on the nipple, it must be inserted deeper into its mouth, so that the front part of the nipple is in contact with the palate and the upper surface of the tongue. Only this stimulus will release the jaw, tongue and cheek action of intense sucking. To achieve this juxtaposition, the region of breast immediately behind the nipple must be pliable and yielding. It is the length of 'hold' that the baby can manage on this yielding tissue which is critical. It is essential that suckling should be fully operative within four or five days of birth, if the breast-feeding process is to be successfully developed. If repeated failure occurs during the first week, the infant will never give the full response. It will have become fixated on the more rewarding (bottle) alternative offered.

Another suckling difficulty is the so-called 'fighting at the breast' response of certain infants. This often gives the mother the impression that the baby does not want to suck, but in reality it means that, despite desperate attempts to do so, it is failing because it is being suffocated. A slightly maladjusted posture of the baby's head at the breast will block the nose and, with the mouth full, there is no way for it to breathe. It is fighting, not to avoid sucking, but for air. There are, of course, many such problems that face the new mother, but I have selected these two because they seem to add supporting evidence for the idea of the female breast as predominantly a sexual signalling device, rather than an expanded milk machine. It is the solid, rounded shape that causes both these problems. One has only to look at the design of the teats on babies' bottles to see the kind of shape that works best. It is much longer and does not swell out into the great rounded hemisphere that causes so much difficulty for the baby's mouth and nose. It is much closer in design to the feeding apparatus of the female chimpanzee. She develops slightly swollen breasts, but even in full lactation she is flat-chested when compared with the average female of our own species. Her nipples, on the other hand, are much more elongated and protrusive and the infant has little or no difficulty in initiating the sucking activity. Because our females have rather a heavy suckling burden and because the breasts are so obviously a part of the feeding apparatus, we have automatically assumed that their protruding, rounded shape must also be part and parcel of the same parental activity. But it now looks as though this assumption has been wrong and that, for our species, breast design is primarily sexual rather than maternal in function.

Leaving the question of feeding, it is worth looking at one or two aspects of the way a mother behaves towards her baby at other times. The usual fondling, cuddling and cleaning require little comment, but the position in which she holds the baby against her body when resting is rather revealing. Careful American studies have disclosed the fact that 80 per cent of mothers cradle their infants in their left arms, holding them against the left side of their bodies. If asked to explain the significance of this preference most people reply that it is obviously the result of the predominance of right-handedness in the population. By holding the babies on their left arms, the mothers keep their dominant arm free for manipulations. But a detailed analysis shows that this is not the case. True, there is a slight difference between right-handed and left-handed females, but not enough to provide an adequate explanation. It emerges that 83 per cent of right-handed mothers hold the baby on the left side, but then so do 78 per cent of left-handed mothers. In other words, only 22 per cent of the left-handed mothers have their dominant hands free for action. Clearly there must be some other, less obvious explanation.

The only other clue comes from the fact that the heart is on the left side of the mother's body. Could it be that the sound of her heart-beat is the vital factor? And in what way? Thinking along these lines it was argued that perhaps during its existence inside the body of the mother, the growing embryo becomes fixated ('imprinted') on the sound of the heart-beat. If this is so, then the re-discovery of this familiar sound after birth might have a calming effect on the infant, especially as it has just been thrust into a strange and frighteningly new world outside. If this is so then the mother, either instinctively or by an unconscious series of trials and errors, would soon arrive at the discovery that her baby is more at peace if held on the left, against her heart, than on the right.

This may sound far-fetched, but tests have now been carried out which reveal that it is nevertheless the true explanation. Groups of new-born babies in a hospital nursery were exposed for a considerable time to the recorded sound of a heart-beat at a standard rate of 72 beats per minute. There were nine babies in each group and it was found that one or more of them was crying for 60 per cent of the time when the sound was not switched on, but that this figure fell to only 38 per cent when the heart-beat recording was thumping away. The heart-beat groups also showed a greater weight gain than the others, although the amount of food taken was the same in both cases. Clearly the beatless groups were burning up a lot more energy as a result of the vigorous actions of their crying.

Another test was done with slightly older infants at bed-time. In some groups the room was silent, in others recorded lullabies were played. In others a ticking metronome was operating at the heart-beat speed of 72 beats per minute. In still others the heart-beat recording itself was played. It was then checked to see which groups fell asleep more quickly. The heart-beat group dropped off in half the time it took for any of the other groups. This not only clinches the idea that the sound of the heart beating is a powerfully calming stimulus, but it also shows that the response is a highly specific one. The metronome imitation will not do — at least, not for young infants.

So it seems fairly certain that this is the explanation of the mother's left-side approach to baby-holding. It is interesting that when 466 Madonna-and-child paintings (dating back over several hundred years) were analysed for this feature, 373 of them showed the baby on the left breast. Here again the figure was at the 80 per cent level. This contrasts with observations of females carrying parcels, where it was found that 50 per cent carried them on the left and 50 per cent on the right.

What other possible results could this heart-beat imprinting have? It may, for example, explain why we insist on locating feelings of love in the heart rather than the head. As the song says: 'You gotta have heart!' It may also explain why mothers rock their babies to lull them to sleep. The rocking motion is carried on at about the same speed as the heart-beat, and once again it probably 'reminds' the infants of the rhythmic sensations they became so familiar with inside the womb, as the great heart of the mother pumped and thumped away above them.

Nor does it stop there. Right into adult life the phenomenon seems to stay with us. We rock with anguish. We rock back and forth on our feet when we are in a state of conflict. The next time you see a lecturer or an after-dinner speaker swaying rhythmically from side to side, check his speed for heart-beat time. His discomfort at having to face an audience leads him to perform the most comforting movements his body can offer in the somewhat limited circumstances; and so he switches on the old familiar beat of the womb.

Wherever you find insecurity, you are liable to find the comforting heart-beat rhythm in one kind of disguise or another. It is no accident that most folk music and dancing has a syncopated rhythm. Here again the sounds and move-ments take the performers back to the safe world of the womb. It is no accident that teenage music has been called 'rock music'. More recently it has adopted an even more revealing name—it is now called 'beat music'. And what are they singing about?: 'My heart is broken', 'You gave your heart to another', or 'My heart belongs to you'.

Fascinating as this subject is, we must not stray too far from the original question of parental behaviour. Up to this point we have been looking at the mother's behaviour towards the child. We have followed her through the dramatic moments of birth, watched her feeding the child, holding it and comfort-ing it. Now we must turn to the baby itself and study it as it grows.

The average weight of a baby at birth is just over seven pounds, which is slightly more than one-twentieth the weight of the average parent. Growth is very rapid during the first two years of life and remains reasonably fast throughout the following four years. At the age of six, however, it slows down considerably. This phase of gradual growth continues until eleven in boys and until ten in girls. Then, at puberty, it puts on another spurt. Rapid growth is seen again from eleven until seventeen in boys and from ten until fifteen in girls. Because of their slightly earlier puberty, girls tend to outstrip boys between the eleventh and fourteenth years, but then the boys pass them again and stay in front from that point on. Body growth tends to end for girls at around nineteen, and for boys much later, at about twenty-five. The first teeth usually appear around the sixth or seventh month, and the full set of milk teeth is usually complete by the end of the second year or the middle of the third. The permanent teeth erupt in the sixth year, but the final molars—the wisdom teeth—do not usually appear until about the nineteenth.

Newborn infants spend a great deal of time sleeping. It is usually claimed that they only awaken for about two hours a day during the first few weeks, but this is not the case. They are sleepy, but not that sleepy. Careful studies have revealed that the average time spent sleeping during the first three days of life is 16·6 hours out of every 24. Individuals varied a great deal, however, the sleepiest averaging 23 hours out of 24, and the most wide-awake a mere 10·5.

During childhood the sleeping-to-waking ratio gradually shrinks until, by the time the adult stage has been reached, the original sixteen-hour average has been reduced to half. Some adults vary considerably from this typical eight-hour average however. Two out of every hundred require only five hours and another two need as much as ten hours. Adult females, incidentally, have an average sleep-period that is slightly longer than that of adult males.

The sixteen-hour quota of daily sleep at birth does not occur in one long nocturnal session, it is broken up into a number of short periods scattered throughout the twenty-

four hours. Even from birth, however, there is a slight
tendency to sleep more at night than in the day. Gradually, as
the weeks pass, one of the nocturnal sleep periods becomes
longer until it dominates the scene. The infant is now taking
a number of brief 'naps' during the day and a single long sleep
at night. This change brings the daily sleep average down to
about fourteen hours at the age of six months. In the months
that follow, the short daily naps become reduced to two—one
in the morning and one in the afternoon. During the second
year the morning nap usually vanishes, bringing the average
sleep figure down to thirteen hours a day. In the fifth year the
afternoon nap disappears as well, reducing the figure still
further to about twelve hours a day. From this point until
puberty there is a further drop of three hours in the daily
sleep requirement, so that, by the age of thirteen, children are
retiring for only nine hours each night. From this point on,
during adolescence, they do not show any difference from the
fully adult pattern and take no more than eight hours on the
average. The final sleeping rhythm, therefore, matches sexual
maturity rather than final physical maturity.

It is interesting that amongst children of pre-school age, the
more intelligent ones tend to sleep less than the dull ones.
After the age of seven this relationship is reversed, the more
intelligent schoolchildren sleeping more than the dull ones.
By this stage it would seem that, instead of learning more by
being more wide-awake for longer, they are being forced to
learn so much that the more responsive ones are worn out by
the end of the day. Amongst adults, by contrast, there appears
to be no relationship between brilliance and the average
amount of sleep.

The time taken to fall asleep in healthy males and females of
all ages averages about twenty minutes. Waking should occur
spontaneously. The need for an artificial awakening device

indicates that there has been insufficient sleep, and the individual will suffer for it with reduced alertness during the waking period that follows.

During its waking periods the newborn infant moves comparatively little. Unlike other primate species its musculature is poorly developed. A young monkey can cling tightly to its mother from the moment of birth onwards. It may even clasp on to her fur with its hands while it is still in the process of being born. In our own species, by contrast, the newborn is helpless and can only make trivial movements of its arms and legs. Not until it is one month old can it, without assistance, raise its chin up off the ground when lying on its front. At two months it can raise its chest off the ground. At three months it can reach towards suspended objects. At four months it can sit up, with support from the mother. At five months it can sit up on the mother's lap and can grasp objects in the hand. At six months it can sit up in a high chair and successfully grasp dangling objects. At seven months it can sit up alone without assistance. At eight months it can stand up with support from the mother. At nine months it can stand up by holding on to furniture. At ten months it can creep along the ground on its hands and knees. At eleven months it can walk when led by the parent's hand. At twelve months it can pull itself up into a standing position with the help of solid objects. At thirteen months it can climb up a set of stairs. At fourteen months it can stand up by itself and without supporting objects to help it. At fifteen months comes the great moment when, at last, it can walk alone by itself, unaided. (These are all, of course, average figures, but they act as a good rough guide to the postural and locomotory rates of development in our species.)

At about the point where the child has started to walk unaided, it also begins to utter its first words—a few simple

ones at first, but soon the vocabulary blossoms out at a
startling rate. By the age of two the average child can speak
nearly 300 words. By three it has tripled this figure. By four
it can manage nearly 1,600 and by five it has achieved 2,100.
This astonishing rate of learning in the field of vocal imitation
is unique to our species and must be considered as one of our
greatest achievements. It is related, as we saw in Chapter One,
to the pressing need for more precise and helpful communication in connection with co-operative hunting activities. There
is nothing like it, nothing even remotely approaching it, in
other closely related living primates. Chimpanzees are, like
us, brilliant at rapid manipulative imitation, but they cannot
manage vocal imitations. One serious and painstaking attempt
was made to train a young chimpanzee to speak, but with
remarkably limited success. The animal was reared in a house
under conditions identical with those for an infant of our own
species. By combining food rewards with manipulations of its
lips, prolonged attempts were made to persuade it to utter
simple words. By the age of two-and-a-half the animal could
say 'mama', 'papa' and 'cup'. Eventually it managed to say
them in the correct contexts, whispering 'cup' when it wanted
a drink of water. The arduous training continued, but by the
age of six (when our own species would be well over the
2,000-word mark) its total vocabulary extended to no more
than seven words.

This difference is a question of brain, not voice. The
chimpanzee has a vocal apparatus that is structurally perfectly
capable of making a wide variety of sounds. There is no
weakness there that can explain its dumb behaviour. The
weakness is centred inside its skull.

Unlike chimpanzees, certain birds have striking powers of
vocal imitation. Parrots, budgerigars, mynah birds, crows,
and various other species can reel off whole sentences without

batting an eyelid, but unfortunately they are too bird-brained to make good use of this ability. They merely copy the complex sequences of sounds they are taught and repeat them automatically in a fixed order and without any reference to outside events. All the same, it is astonishing that chimpanzees, and monkeys for that matter, cannot achieve better things than they do. Even just a few simple, culturally determined, words would be so useful to them in their natural habitats, that it is difficult to understand why they have not evolved.

Returning to our own species again, the basic, instinctive grunts, moans and screams that we share with other primates are not thrown out by our newly-won verbal brilliance. Our inborn sound signals remain, and they retain their important roles. They not only provide the vocal foundation on which we can build our verbal skyscraper, but they also exist in their own right, as species-typical communication devices. Unlike the verbal signals, they emerge without training and they mean the same in all cultures. The scream, the whimper, the laugh, the roar, the moan and the rhythmic crying convey the same messages to everyone everywhere. Like the sounds of other animals, they relate to basic emotional moods and give us an immediate impression of the motivational state of the vocalizer. In the same way we have retained our instinctive expressions, the smile, the grin, the frown, the fixed stare, the panic face and the angry face. These, too, are common to all societies and persist despite the acquisition of many cultural gestures.

It is intriguing to see how these basic species-sounds and species-faces originate during our early development. The rhythmic crying response is (as we know all too well) present from birth. Smiling arrives later, at about five weeks. Laughing and temper tantrums do not appear until the third or fourth month. It is worth taking a closer look at these patterns.

Crying is not only the earliest mood-signal we give, it is also the most basic. Smiling and laughing are unique and rather specialized signals, but crying we share with thousands of other species. Virtually all mammals (not to mention birds) give vent to high-pitched screams, squeaks, shrieks, or squeals when they are frightened or in pain. Amongst the higher mammals, where facial expressions have evolved as visual signalling devices, these messages of alarm are accompanied by characteristic 'fear-faces'. Whether performed by a young animal or an adult, these responses indicate that something is seriously wrong. The juvenile alerts its parents, the adult alerts the other members of its social group.

As infants a number of things make us cry. We cry if we are in pain, if we are hungry, if we are left alone, if we are faced with a strange and unfamiliar stimulus, if we suddenly lose our source of physical support, or if we are thwarted in attaining an urgent goal. These categories boil down to two important factors: physical pain and insecurity. In either case, when the signal is given, it produces (or should produce) protective responses in the parent. If the child is separated from the parent at the time the signal is given, it immediately has the effect of reducing the distance between them until the infant is held and either rocked, patted or stroked. If the child is already in contact with the parent, or if the crying persists after contact is made, then its body is examined for possible sources of pain. The parental response continues until the signal is switched off (and in this respect it differs fundamentally from the smiling and laughing patterns).

The action of crying consists of muscular tension accompanied by a reddening of the head, watering of the eyes, opening of the mouth, pulling back of the lips, exaggerated breathing with intense expirations and, of course, the high-pitched rasping vocalizations. With older infants it also includes running to the parent and clinging.

I have described this pattern in some detail, despite its familiarity, because it is from this that our specialized signals of laughing and smiling have evolved. When someone says 'they laughed until they cried', he is commenting on this relationship, but in evolutionary terms it is the other way round—we cried until we laughed. How did this come about? To start with, it is important to realize how similar crying and laughing are, as response patterns. Their moods are so different that we tend to overlook this. Like crying, laughing involves muscular tension, opening of the mouth, pulling back of the lips, and exaggerated breathing with intense expirations. At high intensities it also includes reddening of the face and watering of the eyes. But the vocalizations are less rasping and not so high-pitched. Above all, they are shorter and follow one another more rapidly. It is as though the long wail of the crying infant has become segmented, chopped up into little pieces, and at the same time has grown smoother and lower.

It appears that the laughing reaction evolved out of the crying one, as a secondary signal, in the following way. I said earlier that crying is present at birth, but laughing does not appear until the third or fourth month. Its arrival coincides with the development of parental recognition. It may be a wise child that knows its own father, but it is a laughing child that knows its own mother. Before it has learnt to identify its mother's face and to distinguish her from other adults, a baby may gurgle and burble, but it does not laugh. What happens when it starts to single out its own mother is that it also begins to grow afraid of other, strange adults. At two months any old face will do, all friendly adults are welcome. But now its fears of the world around it are beginning to mature and anyone unfamiliar is liable to upset it and start it crying. (Later on it will soon learn that certain other adults can also be rewarding and will lose its fear of them, but this is then done selectively on the basis of personal recognition.) As a result of this process of becoming imprinted on the mother, the infant may find itself placed in a strange conflict. If the mother does something that startles it, she gives it two sets of opposing signals. One set says, 'I am your mother – your personal protector; there is nothing to fear,' and the other set says, 'Look out, there's something frightening here.' This conflict could not arise before the mother was known as an individual, because if she had then done something startling, she would simply be the source of a frightening stimulus at that moment and nothing more. But now she can give the double signal: 'There's danger but there's no danger'. Or, to put it another way: 'There may appear to be danger, but because it is coming from me, you do not need to take it seriously'. The outcome of this is that the child gives a response that is half a crying reaction and half a parental-recognition gurgle. The magic combination produces a laugh. (Or, rather, it did, way back in evolution. It has since become fixed and fully developed as a separate, distinct response in its own right.)

So the laugh says, 'I recognize that a danger is not real,' and it conveys this message to the mother. The mother can now play with the baby quite vigorously without making it cry. The earliest causes of laughter in infants are parental games of 'peek-a-boo', hand-clapping, rhythmical knee-dropping, and lifting high. Later, tickling plays a major role, but not until after the sixth month. These are all shock stimuli, but performed by the 'safe' protector. Children soon learn to provoke them – by play-hiding, for example, so that they will experience the 'shock' of discovery, or play-fleeing so that they will be caught.

Laughter therefore becomes a play signal, a sign that the increasingly dramatic inter-actions between the child and the parent can continue and develop. If they become too frightening or painful, then, of course, the reaction can switch over into crying and immediately re-stimulate the protective response. This system enables the child to expand its exploration of its bodily capacities and the physical properties of the world around it.

Other animals also have special play signals, but compared with ours they are unimpressive. The chimpanzee, for instance, has a characteristic play-face, and a soft play-grunt which is the equivalent of our laughter. In origin these signals have the same kind of ambivalence. When greeting, a young chimpanzee protrudes its lips far forward, stretching them to the limit. When frightened, it retracts them, opening its mouth and exposing the teeth. The play-face, being motivated by both feelings of friendly greeting and fear, is a mixture of the two. The jaws open wide, as in fear, but the lips are pulled forward and keep the teeth covered. The soft grunt is halfway between the 'oo-oo-oo' greeting sound and the scream of fear. If play becomes too rough, the lips pull back and the grunt becomes a short, sharp scream. If it becomes too calm, the jaws close and the lips pull forward into the friendly chimpanzee pout. Basically the situation is the same, then, but the soft play-grunt is a puny signal when compared with our own vigorous, full-blooded laughter. As chimpanzees grow, the significance of the play signal dwindles even more, whereas ours expands and acquires still greater importance in everyday life. The naked ape, even as an adult, is a playful ape. It is all part of his exploratory nature. He is constantly pushing things to their limit, trying to startle himself, to shock himself with-

out getting hurt, and then signalling his relief with peals of
infectious laughter.

Laughing *at* someone can also, of course, become a potent
social weapon among older children and adults. It is doubly
insulting because it indicates that he is both frighteningly odd
and at the same time not worth taking seriously. The professional comedian deliberately adopts this social role and is
paid large sums of money by audiences who enjoy the
reassurance of checking their group normality against his
assumed abnormality.

The response of teenagers to their idols is relevant here. As
an audience, they enjoy themselves, not by screaming with
laughter, but screaming with screams. They not only scream,
they also grip their own and one another's bodies, they writhe,
they moan, they cover their faces and they pull at their hair.
These are all the classic signs of intense pain or fear, but they
have become deliberately stylized. Their thresholds have been
artificially lowered. They are no longer cries for help, but
signals to one another in the audience that they are capable of
feeling an emotional response to the sexual idols which is so
powerful that, like all stimuli of unbearably high intensity,
they pass into the realm of pure pain. If a teenage girl found
herself suddenly alone in the presence of one of her idols, it
would never occur to her to scream at him. The screams were
not meant for him, they were meant for the other girls in the
audience. In this way young girls can reassure one another of
their developing emotional responsiveness.

Before leaving the subject of tears and laughter there is one
further mystery to be cleared up. Some mothers suffer agony
from incessantly crying babies during the first three months of
life. Nothing the parents do seems to stem the flood. They
usually conclude that there is something radically, physically
wrong with the infants and try to treat them accordingly.

They are right, of course, there is something physically wrong; but it is probably effect rather than cause. The vital clue comes with the fact that this so-called 'colic' crying ceases, as if by magic, around the third or fourth month of life. It vanishes at just the point where the baby is beginning to be able to identify its mother as a known individual. A comparison of the parental behaviour of mothers with cry-babies and those with quieter infants gives the answer. The former are tentative, nervous and anxious in their dealings with their offspring. The latter are deliberate, calm and serene. The point is that even at this tender age, the baby is acutely aware of differences in tactile 'security' and 'safety', on the one hand, and tactile 'insecurity' and 'alarm' on the other. An agitated mother cannot avoid signalling her agitation to her new-born infant. It signals back to her in the appropriate manner, demanding protection from the cause of the agitation. This only serves to increase the mother's distress, which in turn increases the baby's crying. Eventually the wretched infant cries itself sick and its physical pains are then added to the sum total of its already considerable misery. All that is necessary to break the vicious circle is for the mother to accept the situation and become calm herself. Even if she cannot manage this (and it is almost impossible to fool a baby on this score) the problem corrects itself, as I said, in the third or fourth month of life, because at that stage the baby becomes imprinted on the mother and instinctively begins to respond to her as the 'protector'. She is no longer a disembodied series of agitating stimuli, but a familiar face. If she continues to give agitating stimuli, they are no longer so alarming because they are coming from a known source with a friendly identity. The baby's growing bond with its parent then calms the mother and automatically reduces her anxiety. The 'colic' disappears.

Up to this point I have omitted the question of smiling

because it is an even more specialized response than laughing. Just as laughing is a secondary form of crying, so smiling is a secondary form of laughing. At first sight it may indeed appear to be no more than a low-intensity version of laughing, but it is not as simple as that. It is true that in its mildest form a laugh is indistinguishable from a smile, and this is no doubt how smiling originated, but it is quite clear that during the course of evolution smiling has become emancipated and must now be considered as a separate entity. High-intensity smiling – the giving of a broad grin, a beaming smile – is completely different in function from high-intensity laughing. It has become specialized as a species greeting signal. If we greet someone by smiling at them, they know we are friendly, but if we greet them by laughing at them, they may have reason to doubt it.

Any social contact is at best mildly fear-provoking. The behaviour of the other individual at the moment of meeting is an unknown quantity. Both the smile and the laugh indicate the existence of this fear and its combination with feelings of attraction and acceptance. But when the laugh develops into high intensity, it signals the readiness for further 'startlement', for further exploitation of the danger-with-safety situation. If, on the other hand, the smiling expression of the low-level laugh grows instead into something else – into a broad grin – it signals that the situation is not to be extended in that way. It indicates simply that the initial mood is an end in itself, without any vigorous elaborations. Mutual smiling reassures the smilers that they are both in a slightly apprehensive, but reciprocally attracted, state of mind. Being slightly fearful means being non-aggressive and being non-aggressive means being friendly, and in this way the smile evolves as a friendly attraction device.

Why, if we have needed this signal, have other primates

managed without it? They do, it is true, have friendly gestures of various kinds, but the smile for us is an additional one, and one of tremendous importance in our daily lives, both as infants and as adults. What is it about our own pattern of existence that has brought it so much to the forefront? The answer, it would seem, lies in our famous naked skins. When a young monkey is born it clings tightly to its mother's fur. There it stays, hour in and hour out, day after day. For weeks, or even months, it never leaves the snug protection of its mother's body. Later, when it is venturing away from her for the first time, it can run back to her at a moment's notice and cling on again in an instant. It has its own positive way of ensuring close physical contact. Even if the mother does not welcome this contact (as the infant grows older and heavier), she will have a hard time rejecting it. Anyone who has ever had to act as a foster-mother for a young chimpanzee can testify to this.

When *we* are born we are in a much more hazardous position. Not only are we too weak to cling, but there is nothing to cling to. Robbed of any mechanical means of ensuring close proximity with our mothers, we must rely entirely on maternally stimulating signals. We can scream our heads off to summon parental attention, but having got it we must do something more to maintain it. A young chimpanzee screams for attention just as we do. The mother rushes over and grabs it up. Instantly the baby is clinging again. This is the moment at which we need a clinging-substitute, some kind of signal that will reward the mother and make her want to stay on with us. The signal we use is the smile.

Smiling begins during the first few weeks of life, but to start with it is not directed at anything in particular. By about the fifth week it is being given as a definite reaction to certain stimuli. The baby's eyes can now fixate objects. At first it is most responsive to a pair of eyes staring at it. Even two black spots on a piece of card will do. As the weeks pass, a mouth also becomes necessary. Two black spots with a mouth-line below them are now more efficient at eliciting the response. Soon a widening of the mouth becomes vital, and then the eyes begin to lose their significance as key stimuli. At this stage, around three to four months, the response starts to become more specific. It is narrowed down from any old face to the particular face of the mother. Parental imprinting is taking place.

The astonishing thing about the growth of this reaction is that, at the time when it is developing, the infant is hopeless at discriminating between such things as squares and triangles, or other sharp geometrical shapes. It seems as if there is a special advance in the maturing of the ability to recognize certain rather limited kinds of shapes—those related to human features—while other visual abilities lag behind. This ensures that the infant's vision is going to dwell on the right kind of object. It will avoid becoming imprinted on some near-by inanimate shape.

By the age of seven months the infant is completely imprinted on its mother. Whatever she does now, she will retain her mother-image for her offspring for the rest of its life. Young ducklings achieve this by the act of following the mother, young apes by clinging to her. We develop the vital bond of attachment via the smiling response.

As a visual stimulus the smile has attained its unique configuration principally by the simple act of turning up the mouth-corners. The mouth is opened to some extent and the lips pulled back, as in the face of fear, but by the addition of the curling up of the corners the character of the expression is radically changed. This development has in turn led to the possibility of another and contrasting facial posture—that of the down-turned mouth. By adopting a mouth-line that is the complete opposite of the smile shape, it is possible to signal an anti-smile. Just as laughing evolved out of crying and smiling out of laughing, so the unfriendly face has evolved, by a pendulum swing, from the friendly face.

But there is more to smiling than a mouth-line. As adults we may be able to convey our mood by a mere twist of the lips, but the infant throws much more into the battle. When smiling at full intensity, it also kicks and waves its arms about, stretches its hands out towards the stimulus and moves them about, produces babbling vocalizations, tilts back its head and protrudes its chin, leans its trunk forward or rolls it to one side, and exaggerates its respiration. Its eyes become brighter and may close slightly; wrinkles appear underneath or alongside the eyes and sometimes also on the bridge of the nose; the skin-fold between the sides of the nose and the sides of the mouth becomes more accentuated, and the tongue may be slightly protruded. Of these various elements the body movements seem to indicate a struggle on the infant's part to make contact with the mother. With its clumsy physique, the baby is probably showing us all that remains of the ancestral primate clinging response.

I have been dwelling on the baby's smile, but smiling is, of course, a two-way signal. When the infant smiles at its mother she responds with a similar signal. Each rewards the other and the bond between them tightens in both directions. You may feel that this is an obvious statement, but there can be a catch in it. Some mothers, when feeling agitated, anxious, or cross with the child, try to conceal their mood by forcing a smile. They hope that the counterfeit face will avoid upsetting the infant, but in reality this trick may do more harm than good. I mentioned earlier that it is almost impossible to fool a baby over questions of maternal mood. In the early years of life we

seem to be acutely responsive to subtle signs of parental agitation and parental calm. At the pre-verbal stages, before the
massive machinery of symbolic, cultural communication has
bogged us down, we rely much more on tiny movements,
postural changes and tones of voice than we need to in later
life. Other species are particularly good at this, too. The
astonishing ability of 'Clever Hans', the famous counting
horse, was in fact based on its acuteness in responding to
minute postural changes in his trainer. When asked to do a
sum, Hans would tap his foot the appropriate number of times
and then stop. Even if the trainer left the room and someone
else took over, it still worked, because as the vital number of
taps was reached, the stranger could not help tensing his body
slightly. We all have this ability ourselves, even as adults (it
is used a great deal by fortune-tellers to judge when they are
on the right lines), but in pre-verbal infants it appears to be
especially active. If the mother is making tense and agitated
movements, no matter how concealed, she will communicate
these to her child. If at the same time she gives a strong smile, it
does not fool the infant, it only confuses it. Two conflicting
messages are being transmitted. If this is done a great deal it may
be permanently damaging and cause the child serious difficulties
when making social contacts and adjustments later in life.

Leaving the subject of smiling, we must now turn to a very
different activity. As the months pass, a new pattern of infant
behaviour begins to emerge: aggression arrives on the scene.
Temper tantrums and angry crying begin to differentiate
themselves from the earlier all-purpose crying response. The
baby signals its aggression by a more broken, irregular form
of screaming and by violent striking out with its arms and
legs. It attacks small objects, shakes large ones, spits and spews,
and tries to bite, scratch or strike anything in reach. At first
these activities are rather random and unco-ordinated. The

crying indicates that fear is still present. The aggression has not yet matured to the point of a pure attack: this will come much later when the infant is sure of itself and fully aware of its physical capacities. When it does develop, it, too, has its own special facial signals. These consist of a tight-lipped glare. The lips are pursed into a hard line, with the mouth-corners held forward rather than pulled back. The eyes stare fixedly at the opponent and the eyebrows are lowered in a frown. The fists are clenched. The child has begun to assert itself.

It has been found that this aggressiveness can be increased by raising the density of a group of children. Under crowded conditions the friendly social interactions between members of a group become reduced, and the destructive and aggressive patterns show a marked rise in frequency and intensity. This is significant when one remembers that in other animals fighting is used not only to sort out dominance disputes, but also to increase the spacing-out of the members of a species. We will return to this in Chapter Five.

Apart from protecting, feeding, cleaning and playing with the offspring, the parental duties also include the all-important process of training. As with other species, this is done by a punishment-and-reward system that gradually modifies and adjusts the trial-and-error learning of the young. But, in addition, the offspring will be learning rapidly by imitation — a process that is comparatively poorly developed in most other mammals, but superbly heightened and refined in ourselves. So much of what other animals must laboriously learn for themselves, we acquire quickly by following the example of our parents. The naked ape is a teaching ape. (We are so attuned to this method of learning that we tend to assume that other species benefit in the same way, with the result that we have grossly over-estimated the role that teaching plays in their lives.)

Much of what we do as adults is based on this imitative absorption during our childhood years. Frequently we imagine that we are behaving in a particular way because such behaviour accords with some abstract, lofty code of moral principles, when in reality all we are doing is obeying a deeply ingrained and long 'forgotten' set of purely imitative impressions. It is the unmodifiable obedience to these impressions (along with our carefully concealed instinctive urges) that makes it so hard for societies to change their customs and their 'beliefs'. Even when faced with exciting, brilliantly rational new ideas, based on the application of pure, objective intelligence, the community will still cling to its old home-based habits and prejudices. This is the cross we have to bear if we are going to sail through our vital juvenile 'blotting-paper' phase of rapidly mopping up the accumulated experiences of previous generations. We are forced to take the biased opinions along with the valuable facts.

Luckily we have evolved a powerful antidote to this weakness which is inherent in the imitative learning process. We have a sharpened curiosity, an intensified urge to explore which work against the other tendency and produce a balance that has the potential of fantastic success. Only if a culture becomes too rigid as a result of its slavery to imitative repetition, or too daring and rashly exploratory, will it flounder. Those with a good balance between the two urges will thrive. We can see plenty of examples of the too rigid and too rash cultures around the world today. The small, backward societies, completely dominated by their heavy burden of taboos and ancient customs, are cases of the former. The same societies, when converted and 'aided' by advanced cultures, rapidly become examples of the latter. The sudden overdose of social novelty and exploratory excitement swamps the stabilizing forces of ancestral imitation and tips the scales too far the other way. The result is cultural turmoil and disintegration. Lucky is the society that enjoys the gradual acquisition of a perfect balance between imitation and curiosity, between slavish, unthinking copying and progressive, rational experimentation.

Chapter Four

*

EXPLORATION

All mammals have a strong exploratory urge, but for some it is more crucial than others. It depends largely on how specialized they have become during the course of evolution. If they have put all their evolutionary effort into the perfection of one particular survival trick, they do not need to bother so much about the general complexities of the world around them. So long as the ant-eater has its ants and the koala bear its gum leaves, then they are well satisfied and the living is easy. The non-specialists, on the other hand—the opportunists of the animal world—can never afford to relax. They are never sure where their next meal may be coming from, and they have to know every nook and cranny, test every possibility, and keep a sharp look-out for the lucky chance. They must explore, and keep on exploring. They must investigate, and keep on re-checking. They must have a constantly high level of curiosity.

It is not simply a matter of feeding: self-defence can make the same demands. Porcupines, hedgehogs and skunks can snuffle and stomp about as noisily as they like, heedless of their enemies, but the unarmed mammal must be forever on the alert. It must know the signs of danger and the routes of escape. To survive it must know its home range in every minute detail.

Looked at in this way, it might seem rather inefficient not to specialize. Why should there be any opportunist mammals at all? The answer is that there is a serious snag in the specialist way of life. Everything is fine as long as the special survival device works, but if the environment undergoes a major change the specialist is left stranded. If it has gone to sufficient extremes to outstrip its competitors, the animal will have been forced to make major changes in its genetical make-up, and it will not be able to reverse these quickly enough when the crunch comes. If the gum-tree forests were swept away the koala would perish. If an iron-mouthed killer developed the ability to munch up porcupine quills, the porcupine would become easy prey. For the opportunist the going may always be tough, but the creature will be able to adapt rapidly to any quick-change act that the environment decides to put on. Take away a mongoose's rats and mice and it will switch to eggs and snails. Take away a monkey's fruit and nuts and it will switch to roots and shoots.

Of all the non-specialists, the monkeys and apes are perhaps the most opportunist. As a group, they have specialized in non-specialization. And among the monkeys and apes, the naked ape is the most supreme opportunist of them all. This is just another facet of his neotenous evolution. All young monkeys are inquisitive, but the intensity of their curiosity tends to fade as they become adult. With us, the infantile inquisitiveness is strengthened and stretched out into our mature years. We never stop investigating. We are never satisfied that we know enough to get by. Every question we answer leads on to another question. This has become the greatest survival trick of our species.

The tendency to be attracted by novelty has been called *neophilia* (love of the new), and has been contrasted with *neophobia* (fear of the new). Everything unfamiliar is potentially dangerous. It has to be approached with caution. Perhaps it should be avoided? But if it is avoided, then how shall we ever know anything about it? The neophilic urge must drive us on and keep us interested until the unknown has become the known, until familiarity has bred contempt and, in the process, we have gained valuable experience to be stored away and called upon when needed at a later date. The child does this all the time. So strong is his urge that parental restraint is necessary. But although parents may succeed in guiding curiosity, they can never suppress it. As children grow older their exploratory tendencies sometimes reach alarming proportions and adults can be heard referring to 'a group of youngsters behaving like wild animals'. But the reverse is actually the case. If the adults took the trouble to study the way in which adult wild animals really do behave, they would find that *they* are the wild animals. They are the ones who are trying to limit exploration and who are selling out to the cosiness of sub-human conservativism. Luckily for the species, there are always enough adults who retain their juvenile inventiveness and curiosity and who enable populations to progress and expand.

When we look at young chimpanzees at play we are immediately struck by the similarity between their behaviour and that of our own children. Both are fascinated by new 'toys'. They fall on them eagerly, lifting them, dropping them, twisting them, banging them and taking them to pieces. They both invent simple games. The intensity of their interest is as strong as ours, and during the first few years of life they do just as well—better, in fact, because their muscle system develops quicker. But after a while they begin to lose ground. Their brains are not complex enough to build on this good beginning. Their powers of concentration are weak and do not grow as their bodies grow. Above all, they lack the ability to communicate in detail with their parents about the inventive techniques they are discovering.

The best way to clarify this difference is to take a specific example. Picture-making, or graphic exploration, is an obvious choice. As a pattern of behaviour it has been vitally important to our species for thousands of years, and we have the prehistoric remnants at Altamira and Lascaux to prove it.

Given the opportunity and suitable materials, young chimpanzees are as excited as we are to explore the visual possibilities of making marks on a blank sheet of paper. The start of this interest has something to do with the investigation-reward principle of obtaining disproportionately large results from the expenditure of comparatively little energy. This can be seen operating in all kinds of play situations. A great deal of exaggerated effort may be put into the activities, but it is those actions that produce an unexpectedly increased feed-back that are the most satisfying. We can call this the play principle of 'magnified reward'. Both chimps and children like banging things and it is those objects which produce the loudest noise for the smallest effort that are preferred. Balls that bounce high when only weakly thrown, balloons that shoot across a room when only lightly touched, sand that can be moulded with the mildest pressure, toys on wheels that roll easily along at the gentlest push, these are the things that have maximum play-appeal.

When first faced with a pencil and paper the infant does not find itself in a very promising situation. The best it can do is to tap the pencil on to the surface. But this leads to a pleasant surprise. The tap does something more than simply make a noise, it produces a visual impact as well. Something comes out of the end of the pencil and leaves a mark on the paper. A line is drawn.

It is fascinating to watch this first moment of graphic discovery by a chimpanzee or a child. It stares at the line, intrigued by the unexpected visual bonus its action has brought. After viewing the result for a moment it repeats the experiment. Sure enough, it works the second time, then again, and again. Soon the sheet is covered with scribble lines. As time passes, drawing sessions become more vigorous. Single, tentative lines, placed on the paper one after the other, give way to multiple back-and-forth scribbling. If there is a choice, crayons, chalks and paints are preferred to pencils because they have an even bolder impact, produce an even bigger visual effect, as they sweep across the paper.

The first interest in this activity appears at about one-and-a-half years of age, in both chimps and children. But it is not until after the second birthday that the bold, confident, multiple scribbling really gains momentum. At the age of three the average child moves into a new graphic phase: it starts to simplify its confused scribbling. Out of the exciting chaos it begins to distil basic shapes. It experiments with crosses, then with circles, squares and triangles. Meandering lines are led round the page until they join up with themselves, enclosing a space. A line becomes an outline.

During the months that follow, these simple shapes are combined, one with another, to produce simple abstract patterns. A circle is cut through by a cross, the corners of a square are joined by diagonal lines. This is the vital stage that precedes the very first pictorial representations. In the child this great break-through comes in the second half of the third year, or the beginning of the fourth. In the chimpanzee, it never comes. The young chimp manages to make fan-patterns, crosses and circles, and it can even achieve a 'marked circle', but it can go no further. It is particularly tantalizing that the marked-circle motif is the immediate precursor of the earliest representation produced by the typical child. What happens is that a few lines or spots are placed inside the outline of the

circle and then, as if by magic, a face stares back at the infant
painter. There is a sudden flash of recognition. The phase of
abstract experimentation, of pattern invention, is over. Now
a new goal must be reached: the goal of perfected representation. New faces are made, better faces, with the eyes and
mouth in the right place. Details are added — hair, ears, a nose,
arms and legs. Other images are born — flowers, houses,
animals, boats, cars. These are heights the young chimp can
never, it seems, attain. After the peak has been reached — the
circle made and its inside area marked — the animal continues
to grow but its pictures do not. Perhaps one day a genius
chimp will be found, but it seems unlikely.

For the child, the representational phase of graphic exploration now stretches out before it, but although it is the major
area of discovery, the older abstract patterning influences still
make themselves felt, especially between the ages of five and
eight. During this period particularly attractive paintings are
produced because they are based on the solid grounding of
the abstract-shape phase. The representational images are still
at a very simple stage of differentiation and they combine
appealingly with the confident, well-established shape-and-
pattern arrangements.

The process by which the dot-filled circle grows into an
accurate full-length portrait is an intriguing one. The discovery that it represents a face does not lead to an overnight
success in perfecting the process. This clearly becomes the
dominant aim, but it takes time (more than a decade, in fact).
To start with, the basic features have to be tidied up a little —
circles for eyes, a good strong horizontal line for a mouth,
two dots or a central circle for a nose. Hairs have to fringe
the outer circle. And there things can pause for a while. The
face, after all, is the most vital and compelling part of the
mother, at least in visual terms. After a while, though, further

progress is made. By the simple device of making some of the hairs longer than the rest, it is possible for this face-figure to sprout arms and legs. These in turn can grow fingers and toes. At this point the basic figure-shape is still founded on the pre-representational circle. This is an old friend and he is staying late. Having become a face he has now become a face and body combined. It does not seem to worry the child at this stage that the arms of its drawing are coming out of the side of what appears to be its head. But the circle cannot hold out for ever. Like a cell, it must divide and bud off a lower, second cell. Alternatively the two leg lines must be joined somewhere along their length, but higher than the feet. In one of these two ways, a body can be born. Whichever happens, it leaves the arms high and dry, sticking out of the side of the head. And there they stay for quite some time, before they are brought down into their more correct position, protruding from the top of the body.

It is fascinating to observe these slow steps being taken, one after the other, as the voyage of discovery tirelessly continues. Gradually more and more shapes and combinations are attempted, more diverse images, more complex colours, and more varied textures. Eventually, accurate representation is achieved and precise copies of the outside world can be trapped and preserved on paper. But at that stage the original exploratory nature of the activity becomes submerged beneath the pressing demands of pictorial communication. Earlier painting and drawing, in the young chimp and the young child, had nothing to do with the act of communicating. It was an act of discovery, of invention, of testing the possibilities of graphic variability. It was 'action-painting', not signalling. It required no reward – it was its own reward, it was play for play's sake. However, like so many aspects of childhood play, it soon becomes merged into other adult pursuits. Social communication makes a take-over bid for it and the original inventiveness is lost, the pure thrill of 'taking a line for a walk' is gone. Only in doodles do most adults allow it to re-emerge. (This does not mean that they have become uninventive, merely that the area of invention has moved on into more complex, technological spheres.)

Fortunately for the exploratory art of painting and drawing, much more efficient technical methods of reproducing images of the environment have now been developed. Photography and its offshoots have rendered representational 'information painting' obsolete. This has broken the heavy chains of responsibility that have been the crippling burden of adult art for so long. Painting can now once again explore, this time in a mature adult form. And this, one need hardly mention, is precisely what it is doing today.

I selected this particular example of exploratory behaviour because it reveals very clearly the differences between us and our nearest living relative, the chimpanzee. Similar comparisons could be made in other spheres. One or two deserve brief mention. Exploration of the world of sound can be observed in both species. Vocal invention, as we have already seen, is for some reason virtually absent in the chimpanzee, but 'percussive drumming' plays an important role in its life. Young chimpanzees repeatedly investigate the noise-potentials of acts of thumping, foot-stamping and clapping. As adults they develop this tendency into prolonged social drumming sessions. One animal after another stamps, screams and tears up vegetation, beating on tree-stumps and hollow logs. These communal displays may last for half an hour or more. Their exact function is unknown, but they have the effect of mutually arousing the members of a group. In our own species, drumming is also the most widespread form of musical expression. It begins early, as in the chimpanzee, when children begin to test out the percussive values of objects around them in much the same way. But whereas the adult chimpanzees never manage much more than a simple rhythmic tattoo, we elaborate it into complex poly-rhythms and augment it with vibrating rattles and pitch variations. We also make additional noises by blowing into hollow cavities and scraping or plucking pieces of metal. The screams and hoots of the chimpanzee become in us inventive chants. Our development of complicated musical performances appears, in simpler social groups, to have played much the same role as the drumming and hooting sessions of the chimpanzees, namely, mutual group arousal. Unlike picture-making, it was not an activity pattern that became commandeered for the transmission of detailed information on a major scale. The sending of messages by drumming sequences in certain cultures was an exception to this rule, but by and large music was developed as a communal mood-provoker and synchronizer. Its inventive and exploratory content became stronger and stronger, however, and, freed of any important 'representational' duties, it has become a major area of abstract aesthetic experimentation. (Because of its other prior information commitments, painting has only just caught up with it.)

Dancing has followed much the same course as music and singing. The chimpanzees include many swaying and jigging movements in their drumming rituals and these also accompany the mood-provoking musical performances of our own species. From there, like music, they have been elaborated and expanded into aesthetically complex performances.

Closely related to dancing has been the growth of gymnastics. Rhythmical physical performances are common in the play of both young chimps and young children. They rapidly become stylized, but retain a strong element of variability within the structured patterns they assume. But the physical games of chimpanzees do not grow and mature, they fizzle out. We, on the other hand, explore their possibilities to the full and elaborate them in our adult lives into many complex forms of exercises and sports. Again they are important as communal synchronizing devices, but essentially they are means of maintaining and expanding our exploration of our physical capacities.

Writing, a formalized offshoot of picture-making, and verbalized vocal communication have, of course, been developed as our major means of transmitting and recording information, but they have also been utilized as vehicles for aesthetic exploration on an enormous scale. The intricate elaboration of our ancestral grunts and squeaks into complex symbolic speech has enabled us to sit and 'play' with thoughts in our heads, and to manipulate our (primarily instructional) word sequences to new ends as aesthetic, experimental playthings.

So, in all these spheres—in painting, sculpture, drawing, music, singing, dancing, gymnastics, games, sports, writing and speech—we can carry on to our heart's content, all through our long lives, complex and specialized forms of exploration and experiment. Through elaborate training, both as performers and observers, we can sensitize our responsiveness to the immense exploratory potential that these pursuits have to offer. If we set aside the secondary functions of these activities (the making of money, gaining of status, and so forth), then they all emerge, biologically, either as the extension into adult life of infantile play-patterns, or as the superimposition on to adult information-communication systems of 'play-rules'.

These rules can be stated as follows: (1) you shall investigate the unfamiliar until it has become familiar; (2) you shall impose rhythmic repetition on the familiar; (3) you shall vary this repetition in as many ways as possible; (4) you shall select the most satisfying of these variations and develop these at the expense of others; (5) you shall combine and re-combine these variations one with another; and (6) you shall do all this for its own sake, as an end in itself.

These principles apply from one end of the scale to the other, whether you are considering an infant playing in the sand, or a composer working on a symphony.

The last rule is particularly important. Exploratory be-haviour also plays a role in the basic survival patterns of feeding, fighting, mating and the rest. But there it is confined to the early appetitive phases of the activity sequences and is geared to their special demands. For many species of animals it is no more than this. There is no exploration for explora-tion's sake. But, amongst the higher mammals and to a supreme extent in ourselves, it has become emancipated as a distinct, separate drive. Its function is to provide us with as subtle and complex an awareness of the world around us, and of our own capacities in relation to it, as possible. This aware-ness is not heightened in the specific contexts of the basic survival goals, but in generalized terms. What we acquire in this way can then be applied anywhere, at any time, in any context.

I have omitted the growth of science and technology from this discussion because it has largely been concerned with specific improvements in the methods employed in achieving the basic survival goals, such as fighting (weapons), feeding (agriculture), nest-building (architecture) and comfort (medi-cine). It is interesting, though, that as time has gone by and the technical developments have become more and more interlocked with one another, the pure exploratory urge has also invaded the scientific sphere. Scientific research—the very name 're-search' gives the game away (and I mean game) —operates very much on the play-principles mentioned earlier. In 'pure' research, the scientist uses his imagination in virtually the same way as the artist. He talks of a beautiful experiment rather than of an expedient one. Like the artist, he is concerned with exploration for exploration's sake. If the results of the studies prove to be useful in the context of some specific survival goal, all to the good, but this is secondary.

In all exploratory behaviour, whether artistic or scientific, there is the ever-present battle between the neophilic and neophobic urges. The former drives us on to new experiences, makes us crave for novelty. The latter holds us back, makes us take refuge in the familiar. We are constantly in a state of shifting balance between the conflicting attractions of the exciting new stimulus and the friendly old one. If we lost our neophilia we would stagnate. If we lost our neophobia, we would rush headlong into disaster. This state of conflict does not merely account for the more obvious fluctuations in fashions and fads, in hair-styles and clothing, in furniture and cars; it is also the very basis of our whole cultural progression. We explore and we retrench, we investigate and we stabilize. Step by step we expand our awareness and understanding both of ourselves and of the complex environment we live in.

Before leaving this topic there is one final, special aspect of exploratory behaviour that cannot go unmentioned. It concerns a critical phase of social play during the infantile period. When it is very young, the infant's social play is directed primarily at the parents, but as it grows the emphasis is shifted from them towards other children of the same age. The child becomes a member of a juvenile 'play group'. This is a critical step in its development. As an exploratory involvement it has far-reaching effects on the later life of the individual. Of course, all forms of exploration at this tender age have long-term consequences—the child that fails to explore music or painting will find these subjects difficult as an adult—but person-to-person play contacts are even more critical than the rest. An adult coming to music, say, for the first time, without childhood exploration of the subject behind him, may find it difficult, but not impossible. A child that has been severely sheltered from social contact as a member of a play group, on the other hand, will always find himself badly hampered in his adult social interactions. Experiments with monkeys have revealed that not only does isolation in infancy produce a socially withdrawn adult, but it also creates an anti-sexual and anti-parental individual. Monkeys that were reared in isolation from other youngsters failed to participate in play-group activities when exposed to them later, as older juveniles. Although the isolates were physically healthy and had grown well in their solitary states, they were quite incapable of joining in the general rough and tumble. Instead they crouched, immobile, in the corner of the play-room, usually clasping their bodies tightly with their arms, or covering their eyes. When they matured, again as physically healthy specimens, they showed no interest in sexual partners. If forcibly mated, female isolates produced offspring in the normal way, but then proceeded to treat them as though they were huge parasites crawling on their bodies. They attacked them, drove them away, and either killed them or ignored them.

Similar experiments with young chimpanzees showed that, in this species, with prolonged rehabilitation and special care it was possible to undo, to some extent, this behavioural damage, but, even so, its dangers cannot be over-estimated. In our own species, over-protected children will always suffer in adult social contacts. This is especially important in the case of only children, where the absence of siblings sets them at a serious initial disadvantage. If they do not experience the socializing effects of the rough-and-tumble of the juvenile play groups, they are liable to remain shy and withdrawn for the rest of their lives, find sexual pair-bonding difficult or impossible and, if they do manage to become parents, will make bad ones.

From this it is clear that the rearing process has two distinct phases—an early, inward-turning one and a later, outward-turning one. They are both vitally important and we can learn a great deal about them from monkey behaviour. During the early phase the infant is loved, rewarded and protected by the mother. It comes to understand security. In the later phase it is encouraged to be more outward-going, to participate in social contacts with other juveniles. The mother becomes less loving and restricts her protective acts to moments of serious panic or alarm, when outside dangers threaten the colony. She may now actually punish the growing offspring if it persists in clinging to her hairy apron-strings in the absence of serious panic. It now comes to understand and accept its growing independence.

The situation should be basically the same for offspring of our own species. If either of these basic phases is parentally mis-handled, the child will be in serious trouble in later life. If it has lacked the early security phase, but has been suitably active in the independence phase, it will find making new social contacts easy enough, but will be unable to maintain them or make any real depth of contact. If it has enjoyed great security in the early phase, but has been over-protected later on, it will find making new adult contacts immensely difficult and will tend to cling desperately to old ones.

If we take a closer look at the more extreme cases of social withdrawal, we can witness anti-exploratory behaviour in its most extreme and characteristic form. Severely withdrawn individuals may become socially inactive, but they are far from physically inactive. They become preoccupied with repetitive stereotypes. For hour after hour they rock or sway, nod or shake, twirl or twitch, or clasp and unclasp themselves. They may suck their thumbs, or other parts of their bodies, prod or pinch themselves, make strange and repetitive facial expressions, or tap or roll small objects rhythmically. We all exhibit 'tics' of this sort occasionally, but for them it becomes a major and prolonged form of physical expression. What happens is that they find the environment so threatening, social contacts so frightening and impossible, that they seek comfort and reassurance by super-familiarizing their behaviour. The rhythmic repetition of an act renders it increasingly familiar and 'safe'. Instead of performing a wide variety of heterogeneous activities, the withdrawn individual sticks to the few he knows best. For him the old saying: 'Nothing ventured, nothing gained' has been re-written: 'Nothing ventured, nothing lost.'

I have already mentioned the comforting regressive qualities of the heart-beat rhythm, and this applies here, too. Many of the patterns seem to operate at about heart-beat speed, but even those that do not, still act as 'comforters' by virtue of their super-familiarity achieved from constant repetition. It has been noticed that socially retarded individuals increase their stereotypes when put into a strange room. This fits in with the ideas expressed here. The increased novelty of the environment heightens the neophobic fears, and heavier demands are made on the comforting devices to counteract this.

The more a stereotype is repeated, the more it becomes like an artificially produced, maternal heart-beat. Its 'friendliness' increases and increases until it becomes virtually irreversible. Even if the extreme neophobia causing it can be removed (which is difficult enough), the stereotype may go twitching on.

As I said, socially adjusted individuals also exhibit these 'tics' from time to time. Usually they occur in stress situations and here, too, they act as comforters. We know all the signs. The executive awaiting a vital phone call taps or drums on his desk; the woman in the doctor's waiting-room, clasps and unclasps her fingers around her handbag; the embarrassed child swings its body left and right, left and right; the expectant father paces back and forth; the student in the exam sucks his pencil; the anxious officer strokes his moustache. In moderation these little anti-exploratory devices are useful. They help us to tolerate the anticipated 'novelty overdose'. If used to excess, however, there is always the danger that they will become irreversible and obsessive, and will persist even when not called for.

Stereotypes also crop up in situations of extreme boredom. This can be seen very clearly in zoo animals as well as in our own species. It can sometimes reach frightening proportions. What happens here is that the captive animals would make social contacts if only they had the chance, but they are physically prevented from doing so. The situation is basically the same as in cases of social withdrawal. The restricted environment of the zoo cage blocks their social contacts and forces them into a situation of social withdrawal. The cage-bars are a solid, physical equivalent of the psychological barriers facing the socially withdrawn individual. They constitute a powerful anti-exploratory device and, left with nothing to explore, the zoo animal begins to express itself in the only way possible, by developing rhythmic stereotypes. We are all familiar with the repetitive pacing to-and-fro of the caged animal, but this is only one of the many strange patterns that arise. Stylized masturbation may occur. Sometimes this no longer involves manipulation of the penis. The animal (usually a monkey) simply makes the back and forth masturbatory movements of its arm and hand, but without actually touching the penis. Some female monkeys repeatedly suck their own nipples. Young animals suck their paws. Chimpanzees may prod pieces of straw into their (previously healthy) ears. Elephants nod their heads for hours on end. Certain creatures repeatedly bite themselves, or pull their own hair out. Serious self-mutilation may occur. Some of these responses are given in stressful situations, but many of them are simply reactions to boredom. When there is no variability in the environment the exploratory urge stagnates.

Simply by looking at an isolated animal performing one of these stereotypes it is impossible to know for certain what is causing the behaviour. It may be boredom, or it may be stress. If it is stress it may be the result of the immediate environmental situation, or it may be a long-term phenomenon stemming from an abnormal upbringing. A few simple experiments can give us the answer. When a strange object is placed in the cage, if the stereotypes disappear and exploration begins, then they were obviously being caused by boredom. If the stereotypes increase, however, then they were being caused by stress. If they persist after the introduction of other members of the same species, producing a normal social environment, then the individual with the stereotypes has almost certainly had an abnormally isolated infancy.

All these zoo peculiarities can be seen in our own species (perhaps because we have designed our zoos so much like our cities). They should be a lesson to us, reminding us of the enormous importance of achieving a good balance between our neophobic and neophilic tendencies. If we do not have this, we cannot function properly. Our nervous systems will do the best they can for us, but the results will always be a travesty of our true behavioural potentials.

Chapter Five

*

FIGHTING

If we are to understand the nature of our aggressive urges, we must see them against the background of our animal origins. As a species we are so preoccupied with mass-produced and mass-destroying violence at the present time, that we are apt to lose our objectivity when discussing this subject. It is a fact that the most level-headed intellectuals frequently become violently aggressive when discussing the urgent need to suppress aggression. This is not surprising. We are, to put it mildly, in a mess, and there is a strong chance that we shall have exterminated ourselves by the end of the century. Our only consolation will have to be that, as a species, we have had an exciting term of office. Not a long term, as species go, but an amazingly eventful one. But before we examine our own bizarre perfections of attack and defence, we must examine the basic nature of violence in the spearless, gunless, bombless world of animals.

Animals fight amongst themselves for one of two very good reasons: either to establish their dominance in a social hierarchy, or to establish their territorial rights over a particular piece of ground. Some species are purely hierarchical, with no fixed territories. Some are purely territorial, with no hierarchy problems. Some have hierarchies on their territories and have to contend with both forms of aggression. We belong to the last group: we have it both ways. As primates we were already loaded with the hierarchy system.

This is the basic way of primate life. The group keeps moving about, rarely staying anywhere long enough to establish a fixed territory. Occasional inter-group conflict may arise, but it is weakly organized, spasmodic and of comparatively little importance in the life of the average monkey. The 'peck order' (so-called because it was first discussed in respect of chickens) is, on the other hand, of vital significance in his day-to-day — and even his minute-to-minute — living. There is a rigidly established social hierarchy in most species of monkeys and apes, with a dominant male in charge of the group, and the others ranged below him in varying degrees of subordination. When he becomes too old or weak to maintain his domination, he is overthrown by a younger, sturdier male, who then assumes the mantle of the colony boss. (In some cases the usurper literally assumes the mantle, growing one in the form of a cape of long hair.) As the troop keeps together all the time, his role as group tyrant is incessantly operative. But despite this he is invariably the sleekest, best-groomed and sexiest monkey in the community.

Not all primate species are violently dictatorial in their social organization. There is nearly always a tyrant, but he is sometimes a benign and rather tolerant tyrant, as in the case of the mighty gorilla. He shares the females amongst the lesser males, is generous at feeding times, and only asserts himself when something crops up that cannot be shared, or when there are signs of a revolt, or unruly fighting amongst the weaker members.

This basic system obviously had to be changed when the naked ape became a co-operative hunter with a fixed home base. Just as with sexual behaviour, the typical primate system had to be modified to match his adopted carnivore role. The group had to become territorial. It had to defend the region of its fixed base. Because of the co-operative nature of the hunting, this had to be done on a group basis, rather than individually. Within the group the tyrannical hierarchy system of the usual primate colony had to be modified considerably to ensure full co-operation from the weaker members when out hunting. But it could not be abolished altogether. There had to be a mild hierarchy, with stronger members and a top leader, if firm decisions were going to be taken, even if this leader was obliged to take the feelings of his inferiors more into account than his hairy, forest-dwelling equivalent would have to do.

In addition to group defence of territory and hierarchy organization, the prolonged dependency of the young, forcing us to adopt pair-bonded family units, demanded yet another form of self-assertion. Each male, as the head of a family, became involved in defending his own individual home base inside the general colony base. So for us there are three fundamental forms of aggression, instead of the usual one or two. As we know to our cost, they are all still very much in evidence today, despite the complexities of our societies.

How does the aggression work? What are the patterns of behaviour involved? How do we intimidate one another? We must look again at the other animals. When a mammal becomes aggressively aroused a number of basic physiological changes occur within its body. The whole machine has to gear itself up for action, by means of the autonomic nervous system. This system consists of two opposing and counter-balancing sub-systems – the sympathetic and the parasympathetic. The former is the one that is concerned with preparing the body for violent activity. The latter has the task of preserving and restoring bodily reserves. The former says, 'You are stripped for action, get moving;' the latter says, 'Take it easy, relax and conserve your strength.' Under normal circumstances the body listens to both these voices and maintains a happy balance between them, but when strong aggression is aroused it listens only to the sympathetic system. When this is activated, adrenalin pours into the blood and the whole circulatory system is profoundly affected. The heart beats faster and blood is transferred from the skin and viscera to the muscles and brain. There is an increase in blood pressure. The rate of production of red blood corpuscles is rapidly stepped up. There is a reduction of the time taken for blood to coagulate. In addition there is a cessation in the processes of digesting and storing food. Salivation is restrained. Movements of the stomach, the secretion of gastric juices, and the peristaltic movements of the intestines are all inhibited. Also, the rectum and bladder do not empty as easily as under normal conditions. Stored carbohydrate is rushed out of the liver and floods the blood with sugar. There is a massive increase in respiratory activity. Breathing becomes quicker and deeper. The temperature-regulating mechanisms are activated. The hair stands on end and there is profuse sweating.

All these changes assist in preparing the animal for battle. As if by magic, they instantly banish fatigue and make large amounts of energy available for the anticipated physical struggle for survival. The blood is pumped vigorously to the sites where it is most needed — to the brain, for quick thinking, and to the muscles, for violent action. The rise in blood sugars increases muscular efficiency. The speeding up of coagulation processes means that any blood spilled as a result of injury will clot more quickly and reduce wastage. The stepped-up release of red blood cells from the spleen, in combination with the increased speed of blood circulation, aids the respiratory system to boost the intake of oxygen and the removal of carbon dioxide. The full hair erection exposes the skin to the air and helps to cool the body, as does the outpouring of sweat from the sweat glands. The dangers of over-heating from excessive activity are therefore reduced.

With all the vital systems activated, the animal is ready to launch into the attack, but there is a snag. Out-and-out fighting may lead to a valuable victory, but it may also involve serious damage to the victor. The enemy invariably provokes fear as well as aggression. The aggression drives the animal on, the fear holds it back. An intense state of inner conflict arises. Typically, the animal that is aroused to fight does not go straight into an all-out attack. It begins by threatening to attack. Its inner conflict suspends it, tensed for combat, but not yet ready to begin it. If, in this state, it presents a sufficiently intimidating spectacle for its opponent, and the latter slinks away, then obviously this is preferable. The victory can be won without the shedding of blood. The species is able to settle its disputes without undue damage to its members and obviously benefits tremendously in the process.

Throughout the higher forms of animal life there has been a strong trend in this direction—the direction of ritualized combat. Threat and counter-threat has largely replaced actual physical combat. Full-blooded fighting does, of course, still take place from time to time, but only as a last resort, when aggressive signalling and counter-signalling have failed to settle a dispute. The strength of the outward signs of the physiological changes I have described indicates to the enemy just how violently the aggressive animal is preparing itself for action.

This works extremely well behaviourally, but physiologically it creates something of a problem. The machinery of the body has been geared up for a massive output of work. But the anticipated exertions do not materialize. How does the autonomic nervous system deal with this situation? It has mustered all its troops at the front line, ready for action, but their very presence has won the war. What happens now?

If physical combat followed on naturally from the massive activation of the sympathetic nervous system, all the body preparations it had made would be fully utilized. The energy would be burned up and eventually the parasympathetic system would reassert itself and gradually restore a state of physiological calm. But in the tense state of conflict between aggression and fear, everything is suspended. The result is that the parasympathetic system fights back wildly and the autonomic pendulum swings frantically back and forth. As the tense moments of threat and counter-threat tick by, we see flashes of parasympathetic activity interspersed with the sympathetic symptoms. Dryness in the mouth may give way to excessive salivation. Tightening of the bowels may collapse and sudden defecation may occur. The urine, held back so strongly in the bladder, may be released in a flood. The removal of blood from the skin may be massively reversed, extreme pallor being replaced by intense flushing and reddening. The deep and rapid respiration may be dramatically interrupted, leading to gasps and sighs. These are desperate attempts on the part of the parasympathetic system to counter-act the apparent extravagance of the sympathetic. Under normal circumstances it would be out of the question for intense reactions in one direction to occur simultaneously with intense reactions in the other, but under the extreme conditions of aggressive threat, everything gets momentarily out of phase. (This explains why, in extreme cases of shock, fainting or swooning can be observed. In such instances the blood that has been rushed to the brain is withdrawn again so violently that it leads to sudden unconsciousness.)

As far as the threat signalling-system is concerned, this physiological turbulence is a gift. It provides an even richer source of signals. During the course of evolution these mood-signs have been built on and elaborated in a number of ways. Defecation and urination have become important territorial scent-marking devices for many species of mammals. The most commonly seen example of this is the way domestic dogs cock their legs against marker-posts in their territories, and the way this activity is increased during threatening encounters between rival dogs. (The streets of our cities are excessively stimulating for this activity because they constitute overlapping territories for so many rivals, and each dog is forced to super-scent these areas in an attempt to compete.) Some species have evolved super-dunging techniques. The hippopotamus has acquired a specially flattened tail that is waggled rapidly back and forth during the act of defecating. The effect is that of shooting dung through a fan, with the result that the faeces are spread out over a wide area. Many species have developed special anal glands that add strong personal scents to the dung.

The circulatory disturbances producing extreme pallor or intense red flushes have become improved as signals by the development of bare patches of skin on the faces of many species and the rumps of others. The gaping and hissing of the respiratory disturbances have been elaborated into grunts and roars and the many other aggressive vocalizations. It has been suggested that this accounts for the origin of the whole communication system of vocal signals. Another basic trend developing out of respiratory turbulence is the evolution of inflation displays. Many species puff themselves up in threat and may inflate specialized air-sacs and pouches. (This is particularly common amongst birds, which already possess a number of air-sacs as a basic part of their respiratory systems.)

Aggressive hair-erection has led to the growth of specialized regions such as crests, capes, manes and fringes. These and other localized hair patches have become highly conspicuous. The hairs have become elongated or stiffened. Their pigmentation has often been drastically modified to produce areas of strong contrast with the surrounding fur. When aggressively aroused, with the hairs standing on end, the animal suddenly appears larger and more frightening, and the display patches become bigger and brighter.

Aggressive sweating has become another source of scent-signals. In many cases there have, once again, been specialized evolutionary trends exploiting this possibility. Certain of the sweat glands have become enormously enlarged as complex scent-glands. These can be found on the faces, feet, tails and various parts of the body of many species.

All these improvements have enriched the communication systems of animals and rendered their mood language more subtle and informative. They make the threatening behaviour of the aroused animal more 'readable' in more precise terms.

But this is only half the story. We have been considering only the autonomic signals. In addition to all these there is another whole range of signals available, which stem from the tensed-up muscular movements and postures of the threatening animal. All that the autonomic system did was to gear the body up ready for muscular action. But what did the muscles do about it? They were stiffened for the onslaught, but no onslaught came. The outcome of this situation is a series of aggressive intention movements, ambivalent actions, and conflict postures. The impulses to attack and to flee pull the body this way and that. It lunges forward, pulls back, twists sideways, crouches down, leaps up, leans in, tilts away. As soon as the urge to attack gets the upper hand, the impulse to flee immediately countermands the order. Every move to withdraw is checked by a move to attack. During the course of evolution this general agitation has become modified into specialized postures of threat and intimidation. The intention movements have become stylized, the ambivalent jerkings have become formalized into rhythmic twistings and shakings. A whole new repertoire of aggressive signals has been developed and perfected.

As a result of this we can witness, in many animal species, elaborate threat rituals and combat 'dances'. The contestants circle one another in a characteristically stilted fashion, their bodies tense and stiff. They may bow, nod, shake, shiver, swing rhythmically from side to side, or make repeated short, stylized runs. They paw the ground, arch their backs, or lower their heads. All these intention movements act as vital communication signals and combine effectively with the autonomic signals to provide a precise picture of the intensity of the aggression that has been aroused, and an exact indication of the balance between the urge to attack and the urge to flee.

But there is yet more to come. There is another important source of special signals, arising from a category of behaviour that has been named displacement activity. One of the side-effects of an intense inner conflict is that an animal sometimes exhibits strange and seemingly irrelevant pieces of behaviour. It is as if the tensed-up creature, unable to perform either of the things it is desperate to do, finds an outlet for its pent-up energy in some other, totally unrelated activity. Its urge to flee blocks its urge to attack and vice-versa, so it vents its feelings in some other way. Threatening rivals can be seen suddenly to perform curiously stilted and incomplete feeding movements, and then return instantly to their full threat postures. Or they may scratch or clean themselves in some way, interspersing these movements with the typical threat manœuvring. Some species perform displacement nest-building actions, picking up pieces of nest material that happen to lie near by and dropping them on to imaginary nests. Others indulge in 'instant sleep', momentarily tucking their heads into a snoozing position, yawning or stretching.

There has been a great deal of controversy about these displacement activities. It has been argued that there is no objective justification for referring to them as irrelevancies. If an animal feeds, it is hungry, and if it scratches it must itch. It is stressed that it is impossible to prove that a threatening animal is not hungry when it performs so-called displacement feeding actions, or that it is not itching when it scratches. But this is armchair criticism, and to anyone who has actually observed and studied aggressive encounters in a wide variety of species, it is patently absurd. The tension and drama of these moments is such that it is ridiculous to suggest that the contestants would break off, even momentarily, to feed for the sake of feeding, or scratch for the sake of scratching, or sleep for the sake of sleeping.

Despite the academic arguments about the causal mechanisms involved in the production of displacement activities, one thing is clear, namely that in functional terms they provide yet one more source for the evolution of valuable threat signals. Many animals have exaggerated these actions in such a way that they have become increasingly conspicuous and showy.

All these activities, then, the autonomic signals, the intention movements, the ambivalent postures and the displacement activities, become ritualized and together provide the animals with a comprehensive repertoire of threat signals. In most encounters they will be sufficient to resolve the dispute without the contestants coming to blows. But if this system fails, as it often does under conditions of extreme crowding, for example, then real fighting follows and the signals give way to the brutal mechanics of physical attack. Then, the teeth are used to bite, slash and stab, the head and

horns to butt and spear, the body to ram, bump and push, the
legs to claw, kick and swipe, the hands to grasp and squeeze,
and sometimes the tail to thrash and whip. Even so, it is
extremely rare for one contestant to kill the other. Species
that have evolved special killing techniques for dealing with
their prey seldom employ these when fighting their own
kind. (Serious errors have sometimes been made in this
connection, with false assumptions about the presumed
relationship between prey-attacking behaviour and rival-
attacking activities. The two are quite distinct in both motivation and performance.) As soon as the enemy has been
sufficiently subdued, it ceases to be a threat and is ignored.
There is no point in wasting additional energy on it, and it is
allowed to slink away without further damage or persecution.

Before relating all these belligerent activities to our own
species, there is one more aspect of animal aggression that
must be examined. It concerns the behaviour of the loser.
When his position has become untenable, the obvious thing
for him to do is to remove himself as fast as he can. But this
is not always possible. His escape route may be physically
obstructed, or, if he is a member of a tightly knit social group,
he may be obliged to stay within range of the victor. In either
of these cases, he must somehow signal to the stronger animal
that he is no longer a threat and that he does not intend to
continue the fight. If he leaves it until he is badly damaged
or physically exhausted, this will become obvious enough, and
the dominant animal will wander off and leave him in peace.
But if he can signal his acceptance of defeat before his position
has deteriorated to this unfortunate extreme, he will be able
to avoid further serious punishment. This is achieved by the
performance of certain characteristic submissive displays.
These appease the attacker and rapidly reduce his aggression,
speeding up the settlement of the dispute.

They operate in several ways. Basically, they either switch off the signals that have been arousing the aggression, or they switch on other, positively non-aggressive signals. The first category simply serve to calm the dominant animal down, the latter help by actively changing his mood into something else. The crudest form of submission is gross inactivity. Because aggression involves violent movement, a static pose automatically signals non-aggression. Frequently this is combined with crouching and cowering. Aggression involves expanding the body to its maximum size, and crouching reverses this and therefore acts as an appeasement. Facing away from the attacker also helps, being the opposite of the posture of frontal attack. Other threat-opposites are also used. If a particular species threatens by lowering its head, then raising the head can become a valuable appeasement gesture. If an attacker erects its hair, then compressing it will serve as a submission device. In certain rare cases a loser will admit defeat by offering a vulnerable area to the attacker. A chimpanzee, for example, will hold out its hand as a gesture of submission, rendering it extremely vulnerable for serious biting. Because an aggressive chimpanzee would never do such a thing, this begging gesture serves to appease the dominant individual.

The second category of appeasement signals operate as re-motivating devices. The subordinate animal sends out signals that stimulate a non-aggressive response and, as this wells up inside the attacker, his urge to fight is suppressed and subdued by it. This is done in one of three main ways. A particularly widespread re-motivator is the adoption of juvenile food-begging postures. The weaker individual crouches and begs from the dominant one in the infantile posture characteristic of the particular species—a device especially favoured by females when they are being attacked by males. It is often so effective that the male responds by regurgitating some food to the female, who then completes the food-begging ritual by swallowing it. Now in a thoroughly paternal, protective mood, the male loses his aggression and the pair calm down together. This is the basis of courtship feeding in many species, especially with birds, where the early stages of pair-formation involve a great deal of aggression on the part of the male. Another re-motivating activity is the adoption of a female sexual posture by the weaker animal. Regardless of its sex, or its sexual condition, it may suddenly assume the female rump-presentation posture. When it displays towards the attacker in this way, it stimulates a sexual response which damps down the mood of aggression. In such situations, a dominant male *or* female will mount and pseudo-copulate with either a submissive male or a submissive female.

A third form of re-motivation involves the arousal of the mood to groom or be groomed. A great deal of social or mutual grooming goes on in the animal world and it is strongly associated with the calmer, more peaceful moments of community life. The weaker animal may either invite the winner to groom it, or may make signals requesting permission to perform the grooming itself. Monkeys make great use of this device and have a special facial gesture to go with it, consisting of rapidly smacking the lips together—a modified, ritualized version of part of the normal grooming ceremony. When one monkey grooms another it repeatedly pops fragments of skin and other detritus into its mouth, smacking its lips as it does so. By exaggerating the smacking movements and speeding them up, it signals its readiness to perform this duty and frequently manages in this way to suppress the aggression of the attacker and persuade it to relax and allow itself to be groomed. After a while the dominant individual is so lulled by this procedure that the weakling can slip away unharmed.

These, then, are the ceremonies and devices by which animals order their aggressive involvements. The phrase 'nature red in tooth and claw' was originally intended to refer to the brutal prey-killing activities of the carnivores, but it has been applied incorrectly in general terms to the whole subject of animal fighting. Nothing could be further from the truth. If a species is to survive, it simply cannot afford to go around slaughtering its own kind. Intra-specific aggression has to be inhibited and controlled, and the more powerful and savage the prey-killing weapons of a particular species are, the stronger must be the inhibitions about using them to settle disputes with rivals. This is the 'law of the jungle' where territorial and hierarchy disagreements are concerned. Those species that failed to obey this law have long since become extinct.

How do we, as a species, measure up to this situation? What is our own special repertoire of threatening and appeasing signals? What are our fighting methods, and how do we control them?

Aggressive arousal produces in us all the same physiological upheavals and muscular tensions and agitations that were described in the general animal context. Like other species, we also show a variety of displacement activities. In some respects we are not as well equipped as other species to develop these basic responses into powerful signals. We cannot intimidate our opponents, for example, by erecting our body hair. We still do it in moments of great shock ('My hair stood on end'), but as a signal it is of little use. In other respects we can do much better. Our very nakedness, which prevents us from bristling effectively, gives us the chance to send powerful flushing and paling signals. We can go 'white with rage', 'red with anger', or 'pale with fear'. It is the white colour we have to watch for here: this spells activity. If it is combined with other actions that signal attack, then it is a vital danger signal. If it is combined with other actions that signal fear, then it is a panic signal. It is caused, you will recall, by the activation of the sympathetic nervous system, the 'go' system, and it is not to be treated lightly. The reddening, on the other hand, is less worrying: it is caused by the frantic counter-balancing attempts of the parasympathetic system, and indicates that the 'go' system is already being undermined. The angry, red-faced opponent who faces you is far less likely to attack than the white-faced, tight-lipped one. Red-face's conflict is such that he is all bottled up and inhibited, but white-face is still ready for action. Neither can be trifled with, but white-face is much more likely to spring in to the attack unless he is immediately appeased or counter-threatened even more strongly.

In a similar vein, rapid deep breathing is a danger signal, but it has already become less of a threat when it develops into irregular snorts and gurgles. The same relationship exists between the dry mouth of incipient attack and the slobbering mouth of the more intensely inhibited assault. Urination, defecation and fainting usually arrive a little later on the scene, following in the wake of the massive shock-wave that accompanies moments of immense tension.

When the urge to attack and escape are both strongly activated simultaneously, we exhibit a number of characteristic intention movements and ambivalent posturings. The most familiar of these is the raising of a clenched fist — a gesture that has become ritualized in two ways. It is performed at some distance from the opponent, at a point where it is too far away to be carried through into a blow. Thus its function is no longer mechanical; instead it has become a visual signal. (With the arm bent and held sideways it has now become the defiant formalized gesture of communist regimes.) It has become further ritualized by the addition of back-and-forth striking movements of the forearm. Fist-shaking of this kind is again visual rather than mechanical in its impact. We perform rhythmically repeated 'blows' with the fist, but still at a safe distance.

While doing this, the whole body may make short approach-intention movements, actions which repeatedly check themselves from going too far. The feet may be stamped forcibly and loudly and the fist brought down and thumped on any near-by object. This last action is an example of something seen frequently in other animals, where it is referred to as a re-direction activity. What happens is that, because the object (the opponent) stimulating the attack is too frightening to be directly assaulted, the aggressive movements are released, but have to be re-directed towards some other, less intimidating object, such as a harmless bystander (we have all suffered from this at one time or another), or even an inanimate object. If the latter is used it may be viciously pulverized or destroyed. When a wife smashes a vase to the floor, it is, of course, really her husband's head that lies there broken into small pieces. It is interesting that chimpanzees and gorillas frequently perform their own versions of this display, when they tear up, smash, and throw around branches and vegetation. Again, it has a powerful visual impact.

A specialized and important accompaniment to all these aggressive displays is the making of threatening facial expressions. These, along with our verbalized vocal signals, provide our most precise method of communicating our exact aggressive mood. Although our smiling face, discussed in an earlier chapter, is unique to our species, our aggressive faces, expressive though they may be, are much the same as those of all the other higher primates. (We can tell a fierce monkey or a scared monkey at a glance, but we have to learn the friendly monkey face.) The rules are quite simple: the more the urge to attack dominates the urge to flee, the more the face pulls itself forwards. When the reverse is the case and fear gets the upper hand, then all the facial details are pulled back. In the attack face, the eyebrows are brought forward in a frown, the forehead is smooth, the mouth-corners are held forward, and the lips make a tight, pursed line. As fear comes to dominate the mood, a scared-threat face appears. The eyebrows are raised, the forehead wrinkles, the mouth-corners are pulled back and the lips part, exposing the teeth. This face often accompanies other gestures that appear to be very aggressive, and such things as forehead-wrinkling and teeth-baring are sometimes thought of as 'fierce' signals because of this. But in fact they are fear signs, the face providing an early-warning signal that fear is very much present, despite the persistence of intimidating gestures by the rest of the body. It is still, of course, a threatening face and cannot be treated smugly. If full fear were being expressed, the face-pulling would be abandoned and the opponent would be retreating.

All this face-making we share with the monkeys, a fact that is worth remembering if ever you come face to face with a large baboon, but there are other faces that we have invented culturally, such as sticking out the tongue, puffing out the cheeks, thumbing the nose, and exaggeratedly screwing up the features, that add considerably to our threat repertoire. Most cultures have also added a variety of threatening or insulting gestures employing the rest of the body. Aggressive intention movements ('hopping mad') have been elaborated into violent war-dances of many different and highly stylized kinds. The function here has become communal arousal and synchronization of strong aggressive feelings, rather than direct visual display to the enemy.

Because, with the cultural development of lethal artificial weapons, we have become such a potentially dangerous species, it is not surprising to find that we have an extra-ordinarily wide range of appeasement signals. We share with the other primates the basic submissive response of crouching and screaming. In addition we have formalized a whole variety of subordinating displays. Crouching itself has become extended into grovelling and prostrating. Minor intensities of it are expressed in the form of kneeling, bowing and curtsying. The key signal here is the lowering of the body in relation to the dominant individual. When threatening, we puff ourselves up to our greatest height, making our bodies as tall and as large as possible. Submissive behaviour must therefore take the opposite course and bring the body down as far as possible. Instead of doing this in a random way, we have stylized it at a number of characteristic, fixed stages, each with its own special signal meaning. The act of saluting is interesting in this context, because it shows how far from the original gesture formalization can carry our cultural signs. At first sight a military salute looks like an aggressive movement. It is similar to the signal version of raising-an-arm-to-strike-a-blow. The vital difference is that the hand is not clenched and it points towards the cap or hat. It is, of course, a stylized modification of the act of removing the hat, which itself was originally part of the procedure of lowering the height of the body.

The distillation of the bowing movement from the original, crude, primate crouch is also interesting. The key feature here is the lowering of the eyes. A direct stare is typical of the most out-and-out aggression. It is part of the fiercest facial expressions and accompanies all the most belligerent gestures. (This is why the children's game of 'stare you out' is so difficult to perform and why the simple curiosity staring of a young child — 'It's rude to stare' — is so condemned.) No matter how reduced in extent the bow becomes by social custom, it always retains the face-lowering element. Male members of a royal court, for example, who, through constant repetition, have modified their bowing reactions, still lower the face, but instead of bending from the waist they now bow stiffly from the neck, lowering only the head region.

On less formal occasions the anti-stare response is given by simple looking-away movements, or a 'shifty-eyed' expression. Only a truly aggressive individual can fix you in the eye for any length of time. During ordinary face-to-face conversations we typically look away from our companions when we are talking, then glance back at them at the end of each sentence, or 'paragraph', to check their response to what we have said. A professional lecturer takes some time to train himself to look directly at the members of his audience, instead of over their heads, down at his rostrum, or out towards the side or back of the hall. Even though he is in such a dominant position, there are so many of them, all staring (from the safety of their seats) at him, that he experiences a basic and initially uncontrollable fear of them. Only after a great deal of practice can he overcome this. The simple, aggressive, physical act of being stared at by a large group of people is also the cause of the fluttering 'butterflies' in the actor's stomach before he makes his entrance on to the stage. He has all his intellectual worries about the qualities of his performance and its reception, of course, but the massed threat-stare is an additional and more fundamental hazard for him. (This is again a case of the curiosity stare being confused at an unconscious level with the threat-stare.) The wearing of spectacles and sunglasses makes the face appear more aggressive because it artificially and accidentally enlarges the pattern of the stare. If we are looked at by someone wearing glasses, we are being given a super-stare. Mild-mannered individuals tend to select thin-rimmed or rimless spectacles (probably without realizing why they do so), because this enables them to see better with the minimum of stare exaggeration. In this way they avoid arousing counter-aggression.

A more intense form of anti-stare is covering the eyes with the hands, or burying the face in the crook of the elbow. The simple act of closing the eyes also cuts off the stare, and it is intriguing that certain individuals compulsively and repeatedly shut their eyes briefly whilst facing and talking to strangers. It is as though their normal blinking responses have become lengthened into extended eye-masking moments. The response vanishes when they are conversing with close friends in a situation where they feel at ease. Whether they are trying to shut off the 'threatening' presence of the stranger, or whether they are attempting to reduce their staring rate, or both, is not always clear.

Because of their powerful intimidating affect, many species have evolved staring eye-spots as self-defence mechanisms. Many moths have a pair of startling eye-markings on their wings. These lie concealed until the creatures are attacked by predators. The wings then open and flash the bright eye-spots in the face of the enemy. It has been proved experimentally that this exerts a valuable intimidating influence on the would-be killers, who frequently flee and leave the insects unmolested. Many fish and some species of birds and even mammals have adopted this technique. In our own species, commercial products have sometimes used the same device (perhaps knowingly, perhaps not). Motor-car designers employ headlamps in this way and frequently add to the overall aggressive impression by sculpturing the line of the front of the bonnet into the shape of a frown. In addition they add 'bared teeth' in the form of a metal grille between the 'eye-spots'. As the roads have become increasingly crowded and driving an increasingly belligerent activity, the threat-faces of cars have become progressively improved and refined, imparting to their drivers a more and more aggressive image. On a smaller scale certain products have been given threat-face brand names, such as OXO, OMO, OZO, and OVO. Fortunately for the manufacturers, these do not repel customers: on the contrary, they catch the eye and, having caught it, reveal themselves to be no more than harmless cardboard boxes. But the impact has already worked, the attention has already been drawn to *that* product rather than to its rivals.

I mentioned earlier that chimpanzees appease by holding out a limp hand towards the dominant individual. We share this gesture with them, in the form of the typical begging or imploring posture. We have also adapted it as a widespread greeting gesture in the shape of the friendly hand-shake. Friendly gestures often grow out of submissive ones. We saw earlier how this happened with the smiling and laughing responses (both of which, incidentally, still appear in appeasing situations as the timid smile and the nervous titter). Hand-shaking occurs as a mutual ceremony between individuals of more or less equal rank, but is transformed into bowing to kiss the held hand when there is strong inequality between the ranks. (With increasing 'equality' between the sexes and the various classes, this latter refinement is now becoming rarer, but still persists in certain specialized spheres where formal dominance hierarchies are rigidly adhered to, as in the case of the Church.) In certain instances hand-shaking has become modified into self-shaking or hand-wringing. In some cultures this is the standard greeting appeasement, in others it is performed only in more extreme 'imploring' contexts.

There are many other cultural specialities in the realm of submissive behaviour, such as throwing in the towel or showing the white flag, but these need not concern us here. One or two of the simpler re-motivating devices do, however, deserve a mention, if only because they bear an interesting relationship to similar patterns in other species. You will recall that certain juvenile, sexual or grooming patterns were performed towards aggressive or potentially aggressive individuals as a method of arousing non-aggressive feelings that competed with the more violent ones and suppressed them. In our own species, juvenile behaviour on the part of submissive adults is particularly common during courtship. The courting pair often adopt 'baby-talk', not because they are heading towards parentalism themselves, but because it arouses tender, protective maternal or paternal feelings in the partner and thereby suppresses more aggressive feelings (or, for that matter, more fearful ones). It is amusing, when thinking back to the development of this pattern into courtship-feeding in birds, to notice the extraordinary increase in mutual feeding that goes on in our own courtship phase. At no other time in our lives do we devote so much effort to popping tasty morsels into one another's mouths, or offering one another boxes of chocolates.

As regards re-motivation in a sexual direction, this occurs wherever a subordinate (male or female) adopts a generalized attitude of 'femininity' towards a dominant individual (male or female) in an aggressive rather than a truly sexual context. This is widespread, but the more specific case of the adoption of the female sexual rump-presentation posture as an appeasement gesture has virtually vanished, along with the disappearance of the original sexual posture itself. It is largely confined now to a form of schoolboy punishment, with rhythmic whipping replacing the rhythmic pelvic thrusts of the dominant male. It is doubtful whether schoolmasters would persist in this practice if they fully appreciated the fact that, in reality, they were performing an ancient primate form of ritual copulation with their pupils. They could just as well inflict pain on their victims without forcing them to adopt the bent-over submissive female posture. (It is significant that schoolgirls are rarely, if ever, beaten in this way – the sexual origins of the act would then become too obvious.) It has been imaginatively suggested by one authority that the reason for sometimes forcing schoolboys to lower their trousers for the administration of the punishment is not related to increasing the pain, but rather to enabling the dominant male to witness the reddening of the buttocks as the beating proceeds, which so vividly recalls the flushing of the primate female hind-quarters when in full sexual condition. Whether this is so or not, one thing is certain about this extraordinary ritual, namely that as a re-motivating appeasement device it is a dismal failure. The more the unfortunate schoolboy stimulates the dominant male crypto-sexually, the more likely he is to per-petuate the ritual and, because the rhythmic pelvic thrusts have become symbolically modified into rhythmic blows of the cane, the victim is right back where he started. He has managed to switch a direct attack into a sexual one, but has then been double-crossed by the symbolic conversion of this sexual one back into another aggressive pattern.

The third re-motivating device, that of grooming, plays a minor, but useful role in our species. We frequently employ stroking and patting movements to soothe an agitated in-dividual, and many of the more dominant members of society spend long hours having themselves groomed and fussed over by subordinates. But we shall return to this subject in another chapter.

Displacement activities also play a part in our aggressive

encounters, appearing in almost any situation of stress or tension. We differ from other animals, however, in that we do not restrict ourselves to a few species-typical displacement patterns. We make use of virtually any trivial actions as outlets for our pent-up feelings. In an agitated state of conflict we may rearrange ornaments, light a cigarette, clean our spectacles, glance at a wrist-watch, pour a drink, or nibble a piece of food. Any of these actions may, of course, be performed for normal functional reasons, but in their displacement activity roles they no longer serve these functions. The ornaments that are rearranged were already adequately displayed. They were not in a muddle and may, indeed, be in a worse state after their agitated rearrangement. The cigarette that is lit in a tense moment may be started when a perfectly good and unfinished one has just been nervously stubbed out. Also, the rate of smoking during tension bears no relation to the physiological addictive nicotine demands of the system. The spectacles that are so laboriously polished are already clean. The watch that is wound up so vigorously does not need winding, and when we glance at it our eyes do not even register what time it tells. When we sip a displacement drink it is not because we are thirsty. When we nibble displacement food it is not because we are hungry. All these actions are performed, not for the normal rewards they bring, but simply for the sake of doing something in an attempt to relieve the tension. They occur with particularly high frequency during the initial stages of social encounters, where hidden fears and aggressions are lurking just below the surface. At a dinner party, or any small social gathering, as soon as the mutual appeasement ceremonies of hand-shaking and smiling are over, displacement cigarettes, displacement drinks and displacement food-snacks are immediately offered. Even at large-scale entertainments such as the theatre and cinema the flow of events is deliberately broken up by short intervals when the audience can indulge in brief bouts of their favourite displacement activities.

When we are in more intense moments of aggressive tension, we tend to revert to displacement activities of a kind that we share with other primate species, and our outlets become more primitive. A chimpanzee in such a situation can be seen to perform repeated and agitated scratching movements, which are of a rather special kind and different from the normal response to an itch. It is confined largely to the head region, or sometimes the arms. The movements themselves are rather stylized. We behave in much the same way, performing stilted displacement grooming actions. We scratch our heads, bite our nails, 'wash' our faces with our hands, tug at our beards or moustaches if we have them, or adjust our coiffure, rub, pick, sniff or blow our noses, stroke our ear-lobes, clean our ear-passages, rub our chins, lick our lips, or rub our hands together in a rinsing action. If moments of great conflict are studied carefully, it can be observed that these activities are all carried out in a ritual fashion without the careful localized adjustments of the true cleaning actions. The displacement head-scratch of one individual may differ markedly from its equivalent in another, but each scratcher develops his own rather fixed and characteristic way of doing it. As real cleaning is not involved, it is of little importance that one region gets all the attention while others are ignored. In any social interaction between a small group of individuals the subordinate members of the group can easily be identified by the higher frequency of their displacement self-grooming activities. The truly dominant individual can be recognized by the almost complete absence of such actions. If the ostensibly dominant member of the group does, in fact, perform a larger number of small displacement activities, then this means that his official dominance is being threatened in some way by the other individuals present.

In discussing all these aggressive and submissive behaviour patterns, it has been assumed that the individuals concerned have been 'telling the truth' and have not been consciously and deliberately modifying their actions to achieve special ends. We 'lie' more with our words than our other communication signals, but even so the phenomenon cannot be overlooked entirely. It is extremely difficult to 'utter' untruths with the kind of behaviour patterns we have been discussing, but not impossible. As I have already mentioned, when parents adopt this procedure towards their young children, it usually fails much more drastically than they realize. Between adults, however, who are much more preoccupied with the verbalized information content of the social interactions, it can be more successful. Unfortunately for the behaviour-liar, he typically lies only with certain selected elements of his total signalling system. Others, which he is not aware of, give the game away. The most successful behaviour-liars are those who, instead of consciously concentrating on modifying specific signals, think themselves into the basic mood they wish to convey and then let the small details take care of themselves. This method is frequently used with great success by professional liars, such as actors and actresses. Their entire working lives are spent performing behavioural lies, a process which can sometimes be extremely damaging to their private lives. Politicians and diplomats are also required to perform an undue amount of behavioural lying, but unlike the actors they are not socially 'licensed to lie', and the resultant guilt feelings tend to interfere with their performances. Also, unlike the actors, they do not undergo prolonged training courses.

Even without professional training, it is possible, with a little effort, and a careful study of the facts presented in this book, to achieve the desired effect. I have deliberately tested this out on one or two occasions, with some degree of success, when dealing with the police. I have reasoned that if there is a strong biological tendency to be appeased by submissive gestures, then this predisposition should be open to manipulation if the proper signals are used. Most drivers, when caught by the police for some minor motoring offence, immediately respond by arguing their innocence, or making excuses of some sort for their behaviour. In doing this they are defending their (mobile) territory and are setting themselves up as territorial rivals. This is the worst possible course of action. It forces the police to counter-attack. If, instead, an attitude of abject submission is adopted, it will become increasingly difficult for the police officer to avoid a sensation of appeasement. A total admission of guilt based on sheer stupidity and inferiority puts the policeman into a position of immediate dominance from which it is difficult for him to attack. Gratitude and admiration must be expressed for the efficiency of his action in stopping you. But words are not enough. The appropriate postures and gestures must be added. Fear and submission in both body posture and facial expression must be clearly demonstrated. Above all, it is essential to get quickly out of the car and move away from it towards the policeman. He must not be allowed to approach you, or you have forced him to go out of his way and thereby threatened him. Furthermore, by staying in the car you are remaining in your own territory. By moving away from it you are automatically weakening your territorial status. In addition to this, the sitting posture inside the car is an inherently dominant one. The power of the seated position is an unusual element in our behaviour. No one may sit if the 'king' is standing. When the 'king' rises, everyone rises. This is a special exception to the general rule about aggressive verticality, which states that increasing submissiveness goes with decreasing posture-height. By leaving the car you therefore shed both your territorial rights and your dominant seated position, and put yourself into a suitably weakened state for the submissive actions that follow. Having stood up, however, it is important not to brace the body erect, but to crouch, lower the head slightly and generally sag. The tone of voice is as important as the words used. Anxious facial expressions and looking away movements are also valuable and a few displacement self-grooming activities can be added for good measure.

Unfortunately, as a driver of a car, one is in a basically aggressive mood of territorial defence, and it is extremely difficult to lie about this mood. It requires either considerable practice, or a good working knowledge of non-verbal behaviour signals. If you are a little short on personal dominance in your ordinary life, the experience, even when consciously and deliberately designed, may be too unpleasant, and it will be preferable to pay the fine.

Although this is a chapter about fighting behaviour, we have so far only dealt with methods of avoiding actual combat. When the situation does finally deteriorate into physical contact, the naked ape — unarmed — behaves in a way that contrasts interestingly with that seen in other primates. For them the teeth are the most important weapons, but for us it is the hands. Where they grab and bite, we grab and squeeze, or strike out with clenched fists. Only in infants or very young children does biting play a significant role in unarmed combat. They, of course, have not yet been able to develop their arm and hand muscles sufficiently to make a great impact with them.

We can witness adult un-armed combat today in a number of highly stylized versions, such as wrestling, judo and boxing, but in its original, unmodified form it is now rare. The moment that serious combat begins, artificial weapons of one sort or another are brought into play. In their crudest form, these are thrown or used as extensions of the fist for delivering heavy blows. Under special circumstances chimpanzees have been able to extend their attacks this far. In conditions of semi-captivity they have been observed to pick up a branch and slam it down hard on to the body of a stuffed leopard, or to tear up clods of earth and hurl them across a water ditch at passers-by. But there is little evidence that they use these methods to any extent in the wild state, and none at all that they use them on one another during disputes between rivals. Nevertheless, they give us a glimpse of the way we probably began, with artificial weapons being developed primarily as a means of defence against other species and for the killing of prey. Their use in intra-specific fighting was almost certainly a secondary trend, but once the weapons were there, they became available for dealing with any emergency, regardless of the context.

The simplest form of artificial weapon is a hard, solid, but unmodified, natural object of wood or stone. By simple improvements in the shapes of these objects, the crude actions of throwing and hitting became augmented with the addition of spearing, slashing, cutting and stabbing movements.

The next great behavioural trend in attacking methods was the extension of the distance between the attacker and his enemy, and it is this step that has nearly been our undoing. Spears can work at a distance, but their range is too limited. Arrows are better, but they lack accuracy. Guns widen the gap dramatically, but bombs dropped from the sky can be delivered at an even greater range, and ground-to-ground rockets can carry the attacker's 'blow' further still. The outcome of this is that the rivals, instead of being defeated, are indiscriminately destroyed. As I explained earlier, the proper business of intra-specific aggression at a biological level is the subduing and not the killing of the enemy. The final stages of destruction of life are avoided because the enemy either flees or submits. In both cases the aggressive encounter is then over: the dispute is settled. But the moment that attacking is done from such a distance that the appeasement signals of the losers cannot be read by the winners, then violent aggression is going to go raging on. It can only be consummated by a direct confrontation with abject submission, or the enemy's headlong flight. Neither of these can be witnessed in the remoteness of modern aggression, and the result is wholesale slaughter on a scale unheard of in any other species.

Aiding and abetting this mayhem is our specially evolved co-operativeness. When we improved this important trait in connection with hunting prey, it served us well, but it has now recoiled upon us. The strong urge towards mutual assistance to which it gave rise has become susceptible to powerful arousal in intra-specific aggressive contexts. Loyalty on the hunt has become loyalty in fighting, and war is born. Ironically, it is the evolution of a deep-seated urge to help our fellows that has been the main cause of all the major horrors of war. It is this that has driven us on and given us our lethal gangs, mobs, hordes and armies. Without it they would lack cohesion and aggression would once again become 'personalized'.

It has been suggested that because we evolved as specialized prey-killers, we automatically became rival-killers, and that there is an inborn urge within us to murder our opponents. The evidence, as I have already explained, is against this. Defeat is what an animal wants, not murder; domination is the goal of aggression, not destruction, and basically we do not seem to differ from other species in this respect. There is no good reason why we should. What has happened, however, is that because of the vicious combination of attack remoteness and group co-operativeness, the original goal has become blurred for the individuals involved in the fighting. They attack now more to support their comrades than to dominate their enemies, and their inherent susceptibility to direct appeasement is given little or no chance to express itself. This unfortunate development may yet prove to be our undoing and lead to the rapid extinction of the species.

Not unnaturally, this dilemma has given rise to a great deal of displacement head-scratching. A favourite solution is massive mutual disarmament; but to be effective this would have to be carried to an almost impossible extreme, one that would ensure that all future fighting was carried out as close-contact combat where the automatic, direct appeasement signals could come into operation again. Another solution is to de-patriotize the members of the different social groups; but this would be working against a fundamental biological feature of our species. As fast as alliances could be forged in one direction, they would be broken in another. The natural tendency to form social in-groups could never be eradicated without a major genetical change in our make-up, and one which would automatically cause our complex social structure to disintegrate.

A third solution is to provide and promote harmless, symbolic substitutes for war; but if these really are harmless they will inevitably only go a very small way towards resolving the real problem. It is worth remembering here that this problem, at a biological level, is one of group territorial defence and, in view of the gross overcrowding of our species, also one of group territorial expansion. No amount of boisterous international football is going to solve this.

A fourth solution is the improvement of intellectual control over aggression. It is argued that, since our intelligence has got us into this mess, it is our intelligence that must get us out. Unhappily, where matters as basic as territorial defence are concerned, our higher brain centres are all too susceptible to the urgings of our lower ones. Intellectual control can help us just so far, but no further. In the last resort it is unreliable, and a single, unreasoned, emotional act can undo all the good it has achieved.

The only sound biological solution to the dilemma is massive de-population, or a rapid spread of the species on to other planets, combined if possible with assistance from all four of the courses of action already mentioned. We already know that if our populations go on increasing at their present terrifying rate, uncontrollable aggressiveness will become dramatically increased. This has been proved conclusively with laboratory experiments. Gross over-crowding will produce social stresses and tensions that will shatter our community organizations long before it starves us to death. It will work directly against improvements in intellectual control and will savagely heighten the likelihood of emotional explosion. Such a development can be prevented only by a marked drop in the breeding rate. Unfortunately there are two serious snags here. As already explained, the family unit —which is still the basic unit of all our societies—is a rearing device. It has evolved into its present, advanced and complex state as a system for producing, protecting and maturing off-spring. If this function is seriously curtailed or temporarily eliminated, the pair-bonding pattern will suffer, and this will bring its own brand of social chaos. If, on the other hand, a selective attempt is made to stem the breeding flood, with certain pairs permitted to breed freely and others prevented from doing so, then this will work against the essential co-operativeness of society.

What it amounts to, in simple numerical terms, is that if all adult members of the population form pairs and breed, they can only afford to produce two offspring per pair if the community is to be maintained at a steady level. Each individual will then, in effect, be replacing him- or herself. Allowing for the fact that a small percentage of the population already fails to mate and breed, and that there will always be a number of premature deaths from accidental injury or other causes, the average family size can, in fact, be slightly larger. Even so, this will put a heavier burden on the pair-bond mechanism. The lighter offspring-load will mean that greater efforts will have to be made in other directions to keep the pair-bonds tightly tied. But this is a much smaller hazard, in the long term, than the alternative of suffocating overcrowding.

To sum up then, the best solution for ensuring world peace is the widespread promotion of contraception or abortion. Abortion is a drastic measure and can involve serious emotional disturbance. Furthermore, once a zygote has been formed by the act of fertilization it constitutes a new individual member of society, and its destruction is, in effect, an act of aggression, which is the very pattern of behaviour that we are attempting to control. Contraception is obviously preferable, and any religious or other 'moralizing' factions that oppose it must face the fact that they are engaged in dangerous war-mongering.

Having brought up the question of religion, it is perhaps worthwhile taking a closer look at this strange pattern of animal behaviour, before going on to deal with other aspects of the aggressive activities of our species. It is not an easy subject to deal with, but as zoologists we must do our best to observe what actually happens rather than listen to what is supposed to be happening. If we do this, we are forced to the conclusion that, in a behavioural sense, religious activities consist of the coming together of large groups of people to perform repeated and prolonged submissive displays to appease a dominant individual. The dominant individual concerned takes many forms in different cultures, but always has the common factor of immense power. Sometimes it takes the shape of an animal from another species, or an idealized version of it. Sometimes it is pictured more as a wise and elderly member of our own species. Sometimes it becomes more abstract and is referred to as simply as 'the state', or in other such terms. The submissive responses to it may consist of closing the eyes, lowering the head, clasping the hands together in a begging gesture, kneeling, kissing the ground, or even extreme prostration, with the frequent accompaniment of wailing or chanting vocalizations. If these submissive actions are successful, the dominant individual is appeased. Because its powers are so great, the appeasement ceremonies have to be performed at regular and frequent intervals, to prevent its anger from rising again. The dominant individual is usually, but not always, referred to as a god.

Since none of these gods exist in a tangible form, why have they been invented? To find the answer to this we have to go right back to our ancestral origins. Before we evolved into co-operative hunters, we must have lived in social groups of the type seen today in other species of apes and monkeys. There, in typical cases, each group is dominated by a single male. He is the boss, the overlord, and every member of the group has to appease him or suffer the consequences. He is also most active in protecting the group from outside hazards and in settling squabbles between lesser members. The whole life of a member of such a group revolves around the dominant animal. His all-powerful role gives him a god-like status. Turning now to our immediate ancestors, it is clear that, with the growth of the co-operative spirit so vital for successful group hunting, the application of the dominant individual's authority had to be severely limited if he was to retain the active, as opposed to passive, loyalty of the other group members. They had to want to help him instead of simply fear him. He had to become more 'one of them'. The old-style monkey tyrant had to go, and in his place there arose a more tolerant, more co-operative naked ape leader. This step was essential for the new type of 'mutual-aid' organization that was evolving, but it gave rise to a problem. The total dominance of the Number 1 member of the group having been replaced by a qualified dominance, he could no longer command unquestioning allegiance. This change in the order of things, vital as it was to the new social system, nevertheless left a gap. From our ancient background there remained a need for an all-powerful figure who could keep the group under control, and the vacancy was filled by the invention of a god. The influence of the invented god-figure could then operate as a force additional to the now more restricted influence of the group leader.

At first sight, it is surprising that religion has been so successful, but its extreme potency is simply a measure of the strength of our fundamental biological tendency, inherited directly from our monkey and ape ancestors, to submit ourselves to an all-powerful, dominant member of the group. Because of this, religion has proved immensely valuable as a device for aiding social cohesion, and it is doubtful whether our species could have progressed far without it, given the unique combination of circumstances of our evolutionary origins. It has led to a number of bizarre by-products, such as a belief in 'another life' where we will at last meet up with the god figures. They were, for reasons already explained, unavoidably detained from joining us in the present life, but this

omission can be corrected in an after-life. In order to facilitate
this, all kinds of strange practices have been developed in
connection with the disposal of our bodies when we die. If
we are going to join our dominant overlords, we must be
well prepared for the occasion and elaborate burial ceremonies
must be performed.

Religion has also given rise to a great deal of unnecessary
suffering and misery, wherever it has become over-formalized
in its application, and whenever the professional 'assistants' of
the god figures have been unable to resist the temptation to
borrow a little of his power and use it themselves. But despite
its chequered history it is a feature of our social life that we
cannot do without. Whenever it becomes unacceptable, it is
quietly, or sometimes violently, rejected, but in no time at
all it is back again in a new form, carefully disguised perhaps,
but containing all the same old basic elements. We simply
have to 'believe in something'. Only a common belief will
cement us together and keep us under control. It could be
argued that, on this basis, any belief will do, so long as it is
powerful enough; but this is not strictly true. It must be
impressive and it must be seen to be impressive. Our communal nature demands the performance of and participation
in elaborate group ritual. Elimination of the 'pomp and
circumstance' will leave a terrible cultural gap and the indoctrination will fail to operate properly at the deep, emotional
level so vital to it. Also, certain types of belief are more wasteful and stultifying than others and can side-track a community
into rigidifying patterns of behaviour that hamper its qualita-
tive development. As a species we are a predominantly intelligent and exploratory animal, and beliefs harnessed to this
fact will be the most beneficial for us. A belief in the validity
of the acquisition of knowledge and a scientific understanding
of the world we live in, the creation and appreciation of

aesthetic phenomena in all their many forms, and the broadening and deepening of our range of experiences in day-to-day living, is rapidly becoming the 'religion' of our time. Experience and understanding are our rather abstract god-figures, and ignorance and stupidity will make them angry. Our schools and universities are our religious training centres, our libraries, museums, art galleries, theatres, concert halls and sports arenas are our places of communal worship. At home we worship with our books, newspapers, magazines, radios and television sets. In a sense, we still believe in an after-life, because part of the reward obtained from our creative works is the feeling that, through them, we will 'live on' after we are dead. Like all religions, this one has its dangers, but if we have to have one, and it seems that we do, then it certainly appears to be the one most suitable for the unique biological qualities of our species. Its adoption by an ever growing majority of the world population can serve as a compensating and reassuring source of optimism to set against the pessimism expressed earlier concerning our immediate future as a surviving species.

Before we embarked on this religious discourse, we had been examining the nature of only one aspect of the organization of aggressiveness in our species, namely the group defence of a territory. But as I explained at the beginning of this chapter, the naked ape is an animal with three distinct social forms of aggression, and we must now consider the other two. They are the territorial defence of the family-unit within the larger group-unit, and the personal, individual maintenance of hierarchy positions.

The spatial defence of the home site of the family unit has remained with us through all our massive architectural advances. Even our largest buildings, when designed as living-quarters, are assiduously divided into repetitive units, one per family. There has been little or no architectural 'division of labour'. Even the introduction of communal eating or drinking buildings, such as restaurants and bars, has not eliminated the inclusion of dining-rooms in the family-unit quarters. Despite all the other advances, the design of our cities and towns is still dominated by our ancient, naked-ape need to divide our groups up into small, discrete, family territories. Where houses have not yet been squashed up into blocks of flats, the defended area is carefully fenced, walled, or hedged off from its neighbours, and the demarcation lines are rigidly respected and adhered to, as in other territorial species.

One of the important features of the family territory is that it must be easily distinguished in some way from all the others. Its separate location gives it a uniqueness, of course, but this is not enough. Its shape and general appearance must make it stand out as an easily identifiable entity, so that it can become the 'personalized' property of the family that lives there. This is something which seems obvious enough, but which has frequently been overlooked or ignored, either as a result of economic pressures, or the lack of biological aware-ness of architects. Endless rows of uniformly repeated, identical houses have been erected in cities and towns all over the world. In the case of blocks of flats the situation is even more acute. The psychological damage done to the territorialism of the families forced by architects, planners and builders to live under these conditions is incalculable. Fortunately, the families concerned can impose territorial uniqueness on their dwellings in other ways. The buildings themselves can be painted different colours. The gardens, where there are any, can be planted and landscaped in individual styles. The insides of the houses or flats can be decorated and filled with ornaments, bric-à-brac and personal belongings in profusion. This is usually explained as being done to make the place 'look nice'.

In fact, it is the exact equivalent to another territorial species depositing its personal scent on a landmark near its den. When you put a name on a door, or hang a painting on a wall, you are, in dog or wolf terms, for example, simply cocking your leg on them and leaving your personal mark there. Obsessive 'collecting' of specialized categories of objects occurs in certain individuals who, for some reason, experience an abnormally strong need to define their home territories in this way.

Bearing this in mind, it is amusing to note the large number of cars that contain small mascots and other personal identification symbols, or to watch the business executive moving into a new office and immediately setting out on his desk his favourite personal pen-tray, paper-weight and perhaps a photograph of his wife. The motor-car and the business office are sub-territories, offshoots of his home base, and it is a great relief to be able to cock his leg on these as well, making them into more familiar, 'owned' spaces.

This leaves us with the question of aggression in relation to the social dominance hierarchy. The individual, as opposed to the places he frequents, must also be defended. His social status must be maintained and, if possible improved, but it must be done cautiously, or he will jeopardize his co-operative contacts. This is where all the subtle aggressive and submissive signalling described earlier comes in to play. Group co-operativeness demands and gets a high degree of conformity in both dress and behaviour, but within the bounds of this conformity there is still great scope for hierarchy competitiveness. Because of these conflicting demands it reaches almost incredible degrees of subtlety. The exact form of the knotting of a tie, the precise arrangement of the exposed section of·a breast-pocket handkerchief, minute distinctions in vocal accent, and other such seemingly trivial characteristics, take on a vital social significance in determining the social standing of the individual. An experienced member of society can read them off at a glance. He would be totally at a loss to do so if suddenly jettisoned into the social hierarchy of New Guinea tribesmen, but in his own culture he is rapidly forced to become an expert. In themselves these tiny differences of dress and habit are utterly meaningless, but in relation to the game of juggling for position and holding it in the dominance hierarchy they are all-important.

We did not evolve, of course, to live in huge conglomerations of thousands of individuals. Our behaviour is designed to operate in small tribal groups probably numbering well under a hundred individuals. In such situations every member of the tribe will be known personally to every other member, as is the case with other species of apes and monkeys today. In this type of social organization it is easy enough for the dominance hierarchy to work itself out and become stabilized, with only gradual changes as members become older and die. In a massive city community the situation is much more stressful. Every day exposes the urbanite to sudden contacts with countless strangers, a situation unheard-of in any other primate species. It is impossible to enter into personal hierarchy relationships with all of them, although this would be the natural tendency. Instead they are allowed to go scurrying by, undominated and undominating. In order to facilitate this lack of social contact, anti-touching behaviour patterns develop. This has already been mentioned when dealing with sexual behaviour, where one sex accidentally touches another, but it applies to more than simply the avoidance of sexual behaviour. It covers the whole range of social-relationship initiation. By carefully avoiding staring at one another, gesturing in one another's direction, signalling in any way, or making physical bodily contact, we manage to survive in an otherwise impossibly overstimulating social situation. If the no-touching rule is broken, we immediately apologize to make it clear that it was purely accidental.

Anti-contact behaviour enables us to keep our number of acquaintances down to the correct level for our species. We do this with remarkable consistency and uniformity. If you require confirmation, take the address or phone books of a hundred widely different types of city-dwellers and count up the number of personal acquaintances listed there. You will find that nearly all of them know well about the same number of individuals, and that this number approximates to what we would think of as a small tribal group. In other words, even in our social encounters we are obeying the basic biological rules of our ancient ancestors.

There will of course be exceptions to this rule – individuals who are professionally concerned with making large numbers of personal contacts, people with behaviour defects that make them abnormally shy or lonely, or people whose special psychological problems render them unable to obtain the expected social rewards from their friends and who try to compensate for this by frantic 'socializing' in all directions. But these types account for only a small proportion of the town and city populations. All the rest happily go about their business in what seems to be a great seething mass of bodies, but which is in reality an incredibly complicated series of interlocking and overlapping tribal groups. How little, how very little, the naked ape has changed since his early, primitive days.

Chapter Six

*

FEEDING

The feeding behaviour of the naked ape appears at first sight to be one of his most variable, opportunistic, and culturally susceptible activities, but even here there are a number of basic biological principles at work. We have already taken a close look at the way his ancestral fruit-picking patterns had to become modified into co-operative prey-killing. We have seen how this led to a number of fundamental changes in his feeding routine. Food-seeking had to become more elaborate and carefully organized. The urge to kill prey had to become partially independent of the urge to eat. Food was taken to a fixed home base for consumption. Greater food preparation had to be carried out. Meals became larger and more spaced out in time. The meat component of the diet became dramatically increased. Food storage and food sharing was practised. Food had to be provided by the males for their family units. Defecation activities had to be controlled and modified.

All these changes were taking place over a very long period of time, and it is significant that, despite the great technological advances of recent years, we are still faithful to them. It would seem that they are rather more than mere cultural devices, to be buffeted this way and that by the whims of fashion. Judging by our present-day behaviour, they must, to some extent at any rate, have become deep-seated biological characteristics of our species.

As we have already noted, the improved food-collecting

techniques of modern agriculture have left the majority of the adult males in our societies without a hunting role. They compensate for this by going out to 'work'. Working has replaced hunting, but has retained many of its basic characteristics. It involves a regular trip from the home base to the 'hunting' grounds. It is a predominantly masculine pursuit, and provides opportunities for male-to-male interaction and group activity. It involves taking risks and planning strategies. The pseudo-hunter speaks of 'making a killing in the City'. He becomes ruthless in his dealings. He is said to be 'bringing home the bacon'.

When the pseudo-hunter is relaxing he goes to all-male 'clubs', from which the females are completely excluded. Younger males tend to form into all-male gangs, often 'predatory' in nature. Throughout the whole range of these organizations, from learned societies, social clubs, fraternities, trade unions, sports clubs, masonic groups, secret societies, right down to teenage gangs, there is a strong emotional feeling of male 'togetherness'. Powerful group loyalties are involved. Badges, uniforms and other identification labels are worn. Initiation ceremonies are invariably carried out with new members. The unisexuality of these groupings must not be confused with homosexuality. They have basically nothing to do with sex. They are all primarily concerned with the male-to-male bond of the ancient co-operative hunting group. The important role they play in the lives of the adult males reveals the persistence of the basic, ancestral urges. If this were not so, the activities they promote could just as well be carried on without the elaborate segregation and ritual, and much of it could be done within the sphere of the family units. Females frequently resent the departure of their males to 'join the boys', reacting to it as though it signified some kind of family disloyalty. But they are wrong to do so.

All they are witnessing is the modern expression of the age-old male-grouping hunting tendency of the species. It is just as basic as the male-female bonding of the naked ape and, indeed, evolved in close conjunction with it. It will always be with us, at least until there has been some new and major genetic change in our make-up.

Although working has largely replaced hunting today, it has not completely eliminated the more primitive forms of expression of this basic urge. Even where there is no economic excuse for participating in the pursuit of animal prey, this activity still persists in a variety of forms. Big-game hunting, stag-hunting, fox-hunting, coursing, falconry, wild-fowling, angling and the hunting-play of children are all contemporary manifestations of the ancient hunting urge.

It has been argued that the true motivation behind these present-day activities has more to do with the defeating of rivals than the hunting down of prey; that the desperate creature at bay represents the most hated member of our own species, the one we would so like to see in the same situation. There is undoubtedly an element of truth in this, at least for some individuals, but when these patterns of activity are viewed as a whole it is clear that it can provide only a partial explanation. The essence of 'sport-hunting' is that the prey should be given a fair chance of escaping. (If the prey is merely a substitute for a hated rival, then why give him any chance at all?) The whole procedure of sport-hunting involves a deliberately contrived inefficiency, a self-imposed handicap, on the part of the hunters. They could easily use machine-guns, or more deadly weapons, but that would not be 'playing the game' — the hunting game. It is the challenge that counts, the complexities of the chase and the subtle manœuvres that provide the rewards.

One of the essential features of the hunt is that it is a tremendous gamble and so it is not surprising that gambling, in the many stylized forms it takes today, should have such a strong appeal for us. Like primitive hunting and sport-hunting, it is predominantly a male pursuit and, like them, it is surrounded by seriously observed social rules and rituals.

An examination of our class structure reveals that both sport-hunting and gambling are more the concern of the lower and upper social classes than of the middle classes, and there is a very good reason for this if we accept them as expressions of a basic hunting drive. I pointed out earlier that work has become the major substitute for primitive hunting, but as such it has most benefited the middle classes. For the average lower-class male, the nature of the work he is required to do is poorly suited to the demands of the hunting drive. It is too repetitive, too predictable. It lacks the elements of challenge, luck and risk so essential to the hunting male. For this reason, lower-class males share with the (non-working) upper-class males a greater need to express their hunting urges than do the middle classes, the nature of whose work is much more suited to its role as a hunting substitute.

Leaving hunting and turning now to the next act in the general feeding pattern, we come to the moment of the kill. This element can find a certain degree of expression in the substitute activities of work, sport-hunting and gambling. In sport-hunting the action of killing still occurs in its original form, but in working and gambling contexts it is transformed into moments of symbolic triumph that lack the violence of the physical act. The urge to kill prey is therefore considerably modified in our present-day way of life. It keeps reappearing with startling regularity in the playful (and not so playful) activities of young boys, but in the adult world it is subjected to powerful cultural suppression.

Two exceptions to this suppression are (to some extent)

condoned. One is the sport-hunting already mentioned, and the other is the spectacle of bull-fighting. Although vast numbers of domesticated animals are slaughtered daily, their killing is normally concealed from the public gaze. With bullfighting the reverse is the case, huge crowds gathering to watch and experience by proxy the acts of violent prey-killing.

Within the formal limits of blood-sports these activities are permitted to continue, but not without protest. Outside these spheres, all forms of cruelty to animals are forbidden and punished. This has not always been the case. A few hundred years ago the torture and killing of 'prey' was regularly staged as a public entertainment in Britain and many other countries. It has since been recognized that participation in violence of this kind is liable to blunt the sensitivities of the individuals concerned towards all forms of blood-letting. It therefore constitutes a potential source of danger in our complex and crowded societies, where territorial and dominance restrictions can build up to an almost unbearable degree, sometimes finding release in a flood of pent-up aggression of abnormal savagery.

We have so far been dealing with the earlier stages of the feeding sequence and their ramifications. After hunting and killing, we come to the meal itself. As typical primates we ought to find ourselves munching away on small, non-stop snacks. But we are not typical primates. Our carnivorous evolution has modified the whole system. The typical carnivore gorges itself on large meals, well spaced out in time, and we clearly fall in with this pattern. The tendency persists even long after the disappearance of the original hunting pressures that demanded it. Today it would be quite easy for us to revert to our old primate ways if we had the inclination to do so. Despite this, we stick to well-defined feeding times, just as though we were still engaged in active prey-hunting. Few, if any, of the millions of naked apes alive today indulge in the typical, scattered feeding routine of the other primates. Even in conditions of plenty, we rarely eat more than three, or at the very most, four times a day. For many people, the pattern involves only one or two large daily meals. It could be argued that this is merely a case of cultural convenience, but there is little evidence to support this. It would be perfectly possible, given the complex organization of food supplies that we now enjoy, to devise an efficient system whereby all food was taken in small snacks, scattered throughout the day. Spreading feeding out in this way could be achieved without any undue loss of efficiency once the cultural pattern became adjusted to it, and it would eliminate the need for the major disruptions in other activities caused by the present 'main meal' system. But, because of our ancient predatory past, it would fail to satisfy our basic biological needs.

It is also relevant to consider the question of why we heat our food and eat it while it is still hot. There are three alternative explanations. One is that it helps to simulate 'prey temperature'. Although we no longer consume freshly killed meat, we nevertheless devour it at much the same temperature as other carnivore species. Their food is hot because it has not yet cooled down: ours is hot because we have re-heated it. Another interpretation is that we have such weak teeth that we are forced to 'tenderize' the meat by cooking it. But this does not explain why we should want to eat it while it is still hot, or why we should heat up many kinds of food that do not require 'tenderizing'. The third explanation is that, by increasing the temperature of the food, we improve its flavour. By adding a complicated range of tasty subsidiaries to the main food objects, we can take this process still further. This relates back, not to our adopted carnivory, but to our more ancient primate past. The foods of typical primates have a much wider variety of flavours than those of carnivores. When a carnivore has gone through its complex sequence of hunting, killing and preparing its food, it behaves much more simply and crudely at the actual crunch. It gobbles; it bolts its food down. Monkeys and apes, on the other hand, are extremely sensitive to the subtleties of varying tastiness in their food morsels. They relish them and keep on moving from one flavour to another. Perhaps, when we heat and spice our meals, we are harking back to this earlier primate fastidiousness. Perhaps this is one way in which we resisted the move towards full-blooded carnivory.

Having raised the question of flavour, there is a misunderstanding that should be cleared up concerning the way we receive these signals. How do we taste what we taste? The surface of the tongue is not smooth, but covered with small projections, called papillae, which carry the taste buds. We each possess approximately 10,000 of these taste buds, but in old age they deteriorate and decrease in number, hence the jaded palate of the elderly gastronome. Surprisingly, we can only respond to four basic tastes. They are: sour, salt, bitter and sweet. When a piece of food is placed on the tongue, we register the proportions of these four properties contained in it, and this blending gives the food its basic flavour. Different areas of the tongue react more strongly to one or other of the four tastes. The tip of the tongue is particularly responsive to salt and sweet, the sides of the tongue to sour and the back of the tongue to bitter. The tongue as a whole can also judge the texture and the temperature of the food, but beyond that it cannot go. All the more subtle and varied 'flavours' that we respond to so sensitively are not, in fact, tasted, but smelt. The odour of the food diffuses up into the nasal cavity, where the olfactory membrane is located. When we remark that a particular dish 'tastes' delicious, we are really saying that it tastes and smells delicious. Ironically, when we are suffering from a heavy head cold and our sense of smell is severely reduced, we say that our food is tasteless. In reality, we are tasting it as clearly as we ever did. It is its lack of odour that is worrying us.

Having made this point, there is one aspect of our true tasting that requires special comment, and that is our undeniably prevalent 'sweet-tooth'. This is something alien to the true carnivore, but typically primate-like. As the natural food of primates becomes riper and more suitable for consumption, it usually becomes sweeter, and monkeys and apes have a strong reaction to anything that is strongly endowed with this taste. Like other primates, we find it hard to resist 'sweets'. Our ape ancestry expresses itself, despite our strong meat-eating tendency, in the seeking out of specially sweetened substances. We favour this basic taste more than the others. We have 'sweet shops', but no 'sour shops'. Typically, when eating a full-scale meal, we *end* the often complex sequence of flavours with some sweet substance, so that this is the taste that lingers on afterwards. More significantly, when we occasionally take small, inter-meal snacks (and thereby revert, to a slight extent, to an ancient, primate scatter-feeding pattern), we nearly always choose primate-sweet food objects, such as candy, chocolate, ice-cream, or sugared drinks.

So powerful is this tendency that it can lead us into difficulties. The point is that there are two elements in a food object that make it attractive to us: its nutritive value and its palatability. In nature, these two factors go hand in hand, but in artificially produced foodstuffs they can be separated, and this can be dangerous. Food objects that are nutritionally almost worthless can be made powerfully attractive simply by adding a large amount of artificial sweetener. If they appeal to our old primate weakness by tasting 'super-sweet', we will lap them up and so stuff ourselves with them that we have little room left for anything else: thus the balance of our diet can be upset. This applies especially in the case of growing children. In an earlier chapter I mentioned recent research which has shown that the preference for sweet and fruity odours falls off dramatically at puberty, when there is a shift in favour of flowery, oily and musky odours. The juvenile weakness for sweetness can be easily exploited, and frequently is.

Adults face another danger. Because their food is in general made so tasty—so much more tasty than it would be in nature—its palatability value rises sharply, and eating responses are overstimulated. The result is in many cases an unhealthily overweight condition. To counteract this, all kinds of bizarre 'dieting' regimes are invented. The 'patients' are told to eat this or that, cut down on this or on that, or to exercise in various ways. Unfortunately there is only one true answer to the problem: to eat less. It works like a charm, but since the subject remains surrounded by super-palatability signals, it is difficult for him, or her, to maintain this course of action for any length of time. The overweight individual is also bedevilled by a further complication. I mentioned earlier the phenomenon of 'displacement activities'—trivial, irrelevant actions performed as tension-relievers in moments of stress. As we saw, a very frequent and common form of displacement activity is 'displacement feeding'. In tense moments we nibble small morsels of food or sip unneeded drinks. This may help to relax the tension in us, but it also helps us to put on weight, especially as the 'trivial' nature of the displacement feeding action usually means that we select for the purpose something sweet. If practised repeatedly over a long period, this leads to the well-known condition of 'fat anxiety', and we can witness the gradual emergence of the familiar, rounded contours of guilt-edged insecurity. For such a person, slimming routines will work only if accompanied by other behavioural changes that reduce the initial state of tension. The role of chewing-gum deserves a mention in this context. This substance appears to have developed exclusively as a displacement feeding device. It provides the necessary tension-relieving 'occupational' element, without contributing damagingly to the overall food intake.

Turning now to the variety of foodstuffs eaten by a present-day group of naked apes, we find that the range is extensive. By and large, primates tend to have a wider range of food objects in their diets than carnivores. The latter have become food specialists, whereas the former are opportunists. Careful field studies of a wild population of Japanese macaque monkeys, for example, have revealed that they consume as many as 119 species of plants, in the shape of buds, shoots, leaves, fruits, roots and barks, not to mention a wide variety of spiders, beetles, butterflies, ants and eggs. A typical carnivore's diet is more nutritious, but also much more monotonous.

When we became killers, we had the best of both worlds. We added meat with a high nutritive value to our diet, but we did not abandon our old primate omnivory. During recent times—that is, during the last few thousand years—food-obtaining techniques have improved considerably, but the basic position remains the same. As far as we can tell, the earliest agricultural systems were of a kind that can loosely be described as 'mixed farming'. The domestication of animals and plants advanced side by side. Even today, with our now immensely powerful dominance over our zoological and botanical environments, we still keep both strings to our bow. What has stopped us from swinging further in one direction or the other? The answer seems to be that, with vastly increasing population densities, an all-out reliance on meat would give rise to difficulties in terms of quantity, whereas an exclusive dependence on crops would be dangerous in terms of quality.

It could be argued that, since our primate ancestors had to make do without a major meat component in their diets, we should be able to do the same. We were driven to become flesh-eaters only by environmental circumstances, and now that we have the environment under control, with elaborately cultivated crops at our disposal, we might be expected to return to our ancient primate feeding patterns. In essence, this is the vegetarian (or, as one cult calls itself, fruitarian) creed, but it has had remarkably little success. The urge to eat meat appears to have become too deep-seated. Given the opportunity to devour flesh, we are loth to relinquish the pattern. In this connection, it is significant that vegetarians seldom explain their chosen diet simply by stating that they prefer it to any other. On the contrary, they construct an elaborate justification for it, involving all kinds of medical inaccuracies and philosophical inconsistencies.

Those individuals who are vegetarian by choice ensure a balanced diet by utilizing a wide variety of plant substances, like the typical primates. But for some communities a predominantly meatless diet has become a grim practical necessity rather than an ethical minority-preference. With advancing crop-cultivation techniques and the concentration on a very few staple cereals, a kind of low-grade efficiency has proliferated in certain cultures. The large-scale agricultural operations have permitted the growth of big populations, but their dependency on a few basic cereals has led to serious malnutrition. Such people may breed in large numbers, but they produce poor physical specimens. They survive, but only just. In the same way that abuse of culturally developed weapons can lead to aggressive disaster, abuse of culturally developed feeding techniques can lead to nutritional disaster. Societies that have lost the essential food balance in this way may be able to survive, but they will have to overcome the widespread ill-effects of deficiencies in proteins, minerals and vitamins if they are to progress and develop qualitatively. In all the healthiest and most go-ahead societies today, the meat-and-plant diet balance is well maintained and, despite the dramatic changes that have occurred in the methods of obtaining the nutritional supplies, the progressive naked ape of today is still feeding on much the same basic diet as his ancient hunting ancestors. Once again, the transformation is more apparent than real.

Chapter Seven

*

COMFORT

The place where the environment comes into direct contact with an animal—its body surface—receives a great deal of rough treatment during the course of its life. It is astonishing that it survives the wear and tear and lasts so well. It manages to do so because of its wonderful system of tissue replacement and also because animals have evolved a variety of special comfort movements that help to keep it clean. We tend to think of these cleaning actions as comparatively trivial when considered alongside such patterns as feeding, fighting, fleeing and mating, but without them the body could not function efficiently. For some creatures, such as small birds, plumage maintenance is a matter of life and death. If the feathers are allowed to become bedraggled, the bird will be unable to take off fast enough to avoid its predators and will be unable to keep up its high body temperature if conditions become cold. Birds spend many hours bathing, preening, oiling and scratching themselves and carry out this performance in a long and complicated sequence. Mammals are slightly less complex in their comfort patterns, but nevertheless indulge in a great deal of grooming, licking, nibbling, scratching and rubbing. Like feathers, the hair has to be maintained in good order if it is to keep its owner warm. If it becomes clogged and dirty, it will also increase the risk of disease. Skin parasites have to be attacked and reduced in numbers as far as possible. Primates are no exception to this rule.

In the wild state, monkeys and apes can frequently be seen
to groom themselves, systematically working through the
fur, picking out small pieces of dried skin or foreign bodies.
These are usually popped into the mouth and eaten, or at least
tasted. These grooming actions may go on for many minutes,
the animal giving an impression of great concentration. The
grooming bouts may be interspersed with sudden scratchings
or nibblings, directed at specific irritations. Most mammals
only scratch with the back foot, but a monkey or ape can use
either back or front. Its front limbs are ideally suited to the
cleaning tasks. The nimble fingers can run through the fur and
locate specific trouble spots with great accuracy. Compared
with claws and hooves, the primate's hands are precision
'cleaners'. Even so, two hands are better than one, and this
creates something of a problem. The monkey or ape can
manage to bring both its hands into play when dealing with
its legs, flanks, or front, but cannot really get to grips efficiently in this way with its back, or the arms themselves. Also,
lacking a mirror, it cannot see what it is doing when it is
concentrating on the head region. Here, it can use both hands,
but it must work blind. Obviously, the head, back and arms
are going to be less beautifully groomed than the front, sides
and legs, unless something special can be done for them.

The solution is social grooming, the development of a
friendly mutual aid system. This can be seen in a wide range
of both bird and mammal species, but it reaches a peak of
expression amongst the higher primates. Special grooming
invitation signals have been evolved here and social 'cosmetic'
activities are prolonged and intense. When a groomer monkey
approaches a groomee monkey, the former signals its inten-
tions to the latter with a characteristic facial expression. It
performs a rapid lip-smacking movement, often sticking its
tongue out between each smack. The groomee can signal its

acceptance of the groomer's approach by adopting a relaxed posture, perhaps offering a particular region of its body to be groomed. As I explained in an earlier chapter, the lip-smacking action has evolved as a special ritual out of the repeated particle-tasting movements that take place during a bout of fur-cleaning. By speeding them up and making them more exaggerated and rhythmic, it has been possible to convert them into a conspicuous and unmistakable visual signal.

Because social grooming is a co-operative, non-aggressive activity, the lip-smacking pattern has become a friendly signal. If two animals wish to tighten their bond of friendship, they can do so by repeatedly grooming one another, even if the condition of their fur hardly warrants it. Indeed, there seems to be little relationship today between the amount of dirt on the coat, and the amount of mutual grooming that takes place. Social grooming activities appear to have become almost independent of their original stimuli. Although they still have the vital task of keeping the fur clean, their motivation now appears to be more social than cosmetic. By enabling two animals to stay close together in a non-aggressive, co-operative mood they help to tie tighter the inter-personal bonds between the individuals in the troop or colony.

Out of this friendly signalling system have grown two re-motivating devices, one concerned with appeasement and the other with reassurance. If a weak animal is frightened of a stronger one, it can pacify the latter by performing the lip-smacking invitation signal and then proceed to groom its fur. This reduces the aggression of the dominant animal and helps the subordinate one to become accepted. It is permitted to remain 'in the presence' because of services rendered. Conversely, if a dominant animal wishes to calm the fears of a weaker one, it can do so in the same way. By lip-smacking at it, it can underline the fact that it is not aggressive. Despite its dominant aura, it can show that it means no harm. This particular pattern—a reassurance display—is less often seen than the appeasement variety, simply because primate social life requires it less. There is seldom anything that a weak animal has which a dominant might want and could not take by a direct use of aggression. One exception to this can be seen when a dominant but childless female wants to approach and cuddle an infant belonging to another member of the troop. The young monkey is naturally rather frightened by the approach of the stranger and retreats. On such occasions it is possible to observe the large female attempting to re-assure the tiny infant by making the lip-smacking face at it. If this calms the youngster's fears, the female can then fondle it and continues to calm it by gently grooming it.

Clearly, if we turn now to our own species, we might expect to see some manifestation of this basic primate groom-ing tendency, not only as a simple cleaning pattern, but also in a social context. The big difference, of course, is that we no longer have a luxuriant coat of fur to keep clean. When two naked apes meet and wish to reinforce their friendly relationship they must therefore find some kind of substitute for social grooming. If one studies those situations where, in another primate species, one would expect to see mutual grooming, it is intriguing to observe what happens. To start with it is obvious that smiling has replaced lip-smacking. Its origin as a special infantile signal has already been discussed and we have seen how, in the absence of the clinging response, it became necessary for the baby to have some way of attract-ing and pacifying the mother. Extended into adult life, the smile is clearly an excellent 'grooming-invitation' substitute. But, having invited friendly contact, what next? Somehow it has to be maintained. Lip-smacking is reinforced by groom-ing, but what reinforces smiling? True, the smiling response can be repeated and extended in time long after the initial contact, but something else is needed, something more 'occupational'. Some kind of activity, like grooming, has to be borrowed and converted. Simple observations reveal that the plundered source is verbalized vocalization.

The behaviour pattern of talking evolved originally out of the increased need for the co-operative exchange of information. It grew out of the common and widespread animal phenomenon of non-verbal mood vocalization. From the typical, inborn mammalian repertoire of grunts and squeals there developed a more complex series of learnt sound signals. These vocal units and their combinations and re-combinations became the basis of what we can call *information talking*. Unlike the more primitive non-verbal mood signals, this new method of communication enabled our ancestors to refer to objects in the environment and also to the past and the future as well as to the present. To this day, information talking has remained the most important form of vocal communication for our species. But, having evolved, it did not stop there. It acquired additional functions. One of these took the form of *mood talking*. Strictly speaking, this was unnecessary, because the non-verbal mood signals were not lost. We still can and do convey our emotional states by giving vent to ancient primate screams and grunts, but we augment these messages with verbal confirmation of our feelings. A yelp of pain is closely followed by a verbal signal that 'I am hurt'. A roar of anger is accompanied by the message 'I am furious'. Sometimes the non-verbal signal is not performed in its pure state but instead finds expression as a tone of voice. The words 'I am hurt' are whined or screamed. The words 'I am furious' are roared or bellowed. The tone of voice in such cases is so unmodified by learning and so close to the ancient non-verbal mammalian signalling system that even a dog can understand the message, let alone a foreigner from another race of our own species. The actual words used in such instances are almost superfluous. (Try snarling 'good dog', or cooing 'bad dog' at your pet, and you will see what I mean.) At its crudest and most intense level, mood talking is little more than a 'spilling over' of verbalized sound signalling into an area of communication that is already taken care of. Its value lies in the increased possibilities it provides for more subtle and sensitive mood signalling.

A third form of verbalization is *exploratory talking*. This is talking for talking's sake, aesthetic talking, or, if you like, play talking. Just as that other form of information-transmission, picture-making, became used as a medium for aesthetic exploration, so did talking. The poet paralleled the painter. But it is the fourth type of verbalization that we are concerned with in this chapter, the kind that has aptly been described recently as *grooming talking*. This is the meaningless, polite chatter of social occasions, the 'nice weather we are having' or 'have you read any good books lately' form of talking. It is not concerned with the exchange of important ideas or information, nor does it reveal the true mood of the speaker, nor is it aesthetically pleasing. Its function is to reinforce the greeting smile and to maintain the social togetherness. It is our substitute for social grooming. By providing us with a non-aggressive social preoccupation, it enables us to expose ourselves communally to one another over comparatively long periods, in this way enabling valuable group bonds and friendships to grow and become strengthened.

Viewed in this way, it is an amusing game to plot the course of grooming talk during a social encounter. It plays its most dominant role immediately after the initial greeting ritual. It then slowly loses ground, but has another peak of expression as the group breaks up. If the group has come together for purely social reasons, grooming talk may, of course, persist throughout to the complete exclusion of any kind of information, mood or exploratory talk. The cocktail party is a good example of this, and on such occasions 'serious' talking may even be actively suppressed by the host or hostess, who repeatedly intervene to break up long conversations and rotate the mutual-groomers to ensure maximum social contact. In this way, each member of the party is repeatedly thrown back into a state of 'initial contact', where the stimulus for grooming talk will be strongest. If these non-stop social grooming sessions are to be successful, a sufficiently large number of guests must be invited in order to prevent new contacts from running out before the party is over. This explains the mysterious minimum size that is always automatically recognized as essential for gatherings of this kind. Small, informal dinner parties provide a slightly different situation. Here the grooming talk can be observed to wane as the evening progresses and the verbal exchange of serious information and ideas can be seen to gain in dominance as time passes. Just before the party breaks up, however, there is a brief resurgence of grooming talk prior to the final parting ritual. Smiling also reappears at this point, and the social bonding is in this way given a final farewell boost to help carry it over to the next encounter.

If we switch our observations now to the more formal business encounter, where the prime function of the contact is information talking, we can witness a further decline in the dominance of grooming talk, but not necessarily a total eclipse of it. Here its expression is almost entirely confined to the opening and closing moments. Instead of waning slowly, as at the dinner party, it is suppressed rapidly, after a few polite, initial exchanges. It reappears again, as before, in the closing moments of the meeting, once the anticipated moment of parting has been signalled in some way. Because of the strong urge to perform grooming talk, business groups are usually forced to heighten the formalization of their meetings in some way, in order to suppress it. This explains the origin of committee procedure, where formality reaches a pitch rarely encountered on other private social occasions.

Although grooming talk is the most important substitute we have for social grooming, it is not our only outlet for this activity. Our naked skin may not send out very exciting grooming signals, but other more stimulating surfaces are frequently available and are used as substitutes. Fluffy or furry clothing, rugs, or furniture often release a strong grooming response. Pet animals are even more inviting, and few naked apes can resist the temptation to stroke a cat's fur, or scratch a dog behind the ear. The fact that the animal appreciates this social grooming activity provides only part of the reward for the groomer. More important is the outlet the pet animal's body surface gives us for our ancient primate grooming urges.

As far as our own bodies are concerned, we may be naked over most of our surfaces, but in the head region there is still a long and luxuriant growth of hair available for grooming. This receives a great deal of attention—far more than can be explained on a simple hygienic basis—at the hands of the specialist groomers, the barbers and hairdressers. It is not immediately obvious why mutual hairdressing has not become part of our ordinary domestic social gatherings. Why, for instance, have we developed grooming talk as our special substitute for the more typical primate friendship grooming, when we could so easily have concentrated our original grooming efforts in the head region? The explanation appears to lie in the sexual significance of the hair. In its present form the arrangement of the head hair differs strikingly between the two sexes and therefore provides a secondary sexual characteristic. Its sexual associations have inevitably led to its involvement in sexual behaviour patterns, so that stroking or manipulating the hair is now an action too heavily loaded with erotic significance to be permissible as a simple social friendship gesture. If, as a result of this, it is banned from communal gatherings of social acquaintances, it is necessary to find some other outlet for it. Grooming a cat or a sofa may provide an outlet for the urge to groom, but the need to *be* groomed requires a special context. The hairdressing salon is the perfect answer. Here the customer can indulge in the groomee role to his or her heart's content, without any fear of a sexual element creeping in to the proceedings. By making the professional groomers into a separate category, completely dissociated from the 'tribal' acquaintanceship group, the dangers are eliminated. The use of male groomers for males and female groomers for females reduces the dangers still further. Where this is not done, the sexuality of the groomer is reduced in some way. If a female is attended by a male hairdresser, he usually behaves in an effeminate manner, regardless of his true sexual personality. Males are nearly always groomed by male barbers, but if a female masseuse is employed, she is typically rather masculine.

As a pattern of behaviour, hairdressing has three functions. It not only cleans the hair and provides an outlet for social grooming, but it also decorates the groomee. Decoration of the body for sexual, aggressive, or other social purposes is a widespread phenomenon in the case of the naked ape, and it has been discussed under other headings in other chapters. It has no real place in a chapter on comfort behaviour, except that it so often appears to grow out of some kind of grooming activity. Tattooing, shaving and plucking of hair, manicuring, ear-piercing and the more primitive forms of scarification all seem to have their origin in simple grooming actions. But, whereas grooming talk has been borrowed from elsewhere and utilized as a grooming substitute, here the reverse process has taken place and grooming actions have been borrowed and elaborated for other uses. In acquiring a display function, the original comfort actions concerned with skin care have been transformed into what amounts to skin mutilation.

This trend can also be observed in certain captive animals in a zoo. They groom and lick with abnormal intensity until they have plucked bare patches or inflicted small wounds, either on their own bodies or those of companions. Excessive grooming of this kind is caused by conditions of stress or boredom. Similar conditions may well have provoked members of our own species to mutilate their body surfaces, with the already exposed and hairless skin aiding and abetting the process. In our case, however, our inherent opportunism enabled us to exploit this otherwise dangerous and damaging tendency and press it into service as a decorative display device.

Another and more important trend has also developed out of simple skin care, and that is medical care. Other species have made little progress in this direction, but for the naked ape the growth of medical practice out of social grooming behaviour has had an enormous influence on the successful development of the species, especially in more recent times. In our closest relatives, the chimpanzees, we can already witness the beginning of this trend. In addition to the general skin care of mutual grooming, one chimpanzee has been seen to attend to the minor physical disabilities of another. Small sores or wounds are carefully examined and licked clean. Splinters are carefully removed by pinching the companion's skin between two forefingers. In one instance a female chimpanzee with a small cinder in her left eye was seen to approach a male, whimpering and obviously in distress. The male sat down and examined her intently and then proceeded to remove the cinder with great care and precision, gently using the tips of one finger from each hand. This is more than simple grooming. It is the first sign of true co-operative medical care. But for chimpanzees, the incident described is already the peak of its expression. For our own species, with greatly increased intelligence and co-operativeness, specialized grooming of this kind was to be the starting point of a vast technology of mutual physical aid. The medical world today has reached a condition of such complexity that it has become, in social terms, the major expression of our animal comfort behaviour. From coping with minor discomforts it has expanded to deal with major diseases and gross bodily damage. As a biological phenomenon its achievements are unique, but in becoming rational, its irrational elements have been some-what overlooked. In order to understand this, it is essential to distinguish between serious and trivial cases of 'indisposition'. As with any other species, a naked ape can break a leg or become infected with a vicious parasite on a purely accidental or chance basis. But in the case of trivial ailments, all is not what it seems. Minor infections and sicknesses are usually treated rationally, as if they are simply mild versions of serious illnesses, but there is strong evidence to suggest that they are in reality much more related to primitive 'grooming demands'. The medical symptoms reflect a behavioural problem that has taken a physical form, rather than a true physical problem.

Common examples of 'grooming invitation ailments', as we can call them, include coughs, colds, influenza, backache, headache, stomach upsets, skin rashes, sore throats, biliousness, tonsillitis and laryngitis. The condition of the sufferer is not serious, but sufficiently unhealthy to justify increased attention from social companions. The symptoms act in the same way as grooming invitation signals, releasing comfort behaviour from doctors, nurses, chemists, relations and friends. The groomee provokes friendly sympathy and care and this alone is usually enough to cure the illness. The administering of pills and medicines replaces the ancient grooming actions and provides an occupational ritual that sustains the groomee-groomer relationship through this special phase of social interaction. The exact nature of the chemicals prescribed is almost irrelevant and there is little difference at this level between the practices of modern medicine and those of ancient witch-doctoring.

The objection to this interpretation of minor ailments is likely to be based on the observation that real viruses or bacteria can be proved to be present. If they are there and can be shown to be the medical cause of the cold or stomach ache, then why should we seek for a behavioural explanation? The answer is that in any large city, for example, we are all exposed to these common viruses and bacteria all the time, but we only occasionally fall prey to them. Also, certain individuals are much more susceptible than others. Those members of a community who are either very successful or socially well adjusted rarely suffer from 'grooming invitation ailments'. Those that have temporary or long-standing social problems are, by contrast, highly susceptible. The most intriguing aspect of these ailments is the way they are tailored to the special demands of the individual. Supposing an actress, for example, is suffering from social tensions and strains, what happens? She loses her voice, develops laryngitis, so that she is forced to stop work and take a rest. She is comforted and looked after. The tension is resolved (for the time being, at least). If instead she had developed a skin rash on her body, her costume would have covered it and she could have gone on working. The tension would have continued. Compare her situation with that of an all-in wrestler. For him a loss of voice would be useless as a 'grooming invitation ailment', but a skin rash would be ideal, and it is precisely this ailment that wrestlers' doctors find is the muscle-men's most common complaint. In this connection it is amusing that one famous actress, whose reputation relies on her nude appearances in her films, suffers under stress not from laryngitis, but from skin rash. Because, like the wrestlers, her skin exposure is vital, she falls into their ailment category rather than into that of other actresses.

If the need for comfort is intense, then the ailment becomes more intense. The time in our lives when we receive the most elaborate care and protection is when we are infants in our cots. An ailment that is severe enough to put us helplessly to bed, therefore, has the great advantage of recreating for us all the comforting attention of our secure infancy. We may think we are taking a strong dose of medicine, but in reality it is a strong dose of security that we need and that cures us. (This does not imply malingering. There is no need to malinger. The symptoms are real enough. It is the cause that is be-havioural, not the effects.)

We are all to some extent frustrated groomers, as well as groomees, and the satisfaction that can be obtained from caring for the sick is as basic as the cause of the sickness. Some individuals have such a great need to care for others that they may actively promote and prolong sickness in a companion in order to be able to express their grooming urges more fully. This can produce a vicious circle, with the groomer-groomee situation becoming exaggerated out of all propor-tion, to the extent where a chronic invalid demanding (and getting) constant attention is created. If a 'mutual grooming pair' of this type were faced with the behavioural truth concerning their reciprocal conduct, they would hotly deny

it. Nevertheless, it is astonishing what miraculous cures can
sometimes be worked in such instances when a major social
upheaval occurs in the groomer–groomee (nurse-patient)
environment that has been created. Faith-healers have occasionally exploited this situation with startling results, but unfortunately for them many of the cases they encounter have
physical causes as well as physical effects. Also working against
them is the fact that the physical effects of behaviourally
produced 'grooming invitation ailments' can easily create
irreversible body damage if sufficiently prolonged or intense.
Once this has happened, serious, rational medical treatment is
required.

Up to this point I have been concentrating on the social
aspects of comfort behaviour in our species. As we have seen
there have been great developments in that direction, but this
has not excluded or replaced the simpler kinds of self-cleaning
and self-comfort. Like other primates we still scratch ourselves,
rub our eyes, pick our sores, and lick our wounds. We also
share with them a strong tendency to sun-bathe. In addition
we have added a number of specialized cultural patterns, the
most common and widespread of which is washing with
water. This is rare in other primates, although certain species
bathe occasionally, but for us it now plays the major role in
body cleaning in most communities.

Despite its obvious advantages, frequent cleansing with
water nevertheless puts a severe strain on the production of
antiseptic and protective oils and salts by the skin glands, and
to some extent it is bound to make the body surface more
susceptible to diseases. It survives this disadvantage only
because, at the same time that it eliminates the natural oils and
salts, it removes the dirt that is the source of these diseases.

In addition to problems of keeping clean, the general
category of comfort behaviour also includes those patterns of

activity concerned with the task of maintaining a suitable body temperature. Like all mammals and birds, we have evolved a constant, high body temperature, giving us greatly increased physiological efficiency. If we are healthy, our deep body temperature varies no more than 3° Fahrenheit, regardless of the outside temperature. This internal temperature fluctuates with a daily rhythm, the highest level occurring in the late afternoon and the lowest at around 4 a.m. If the external environment becomes too hot or too cold we quickly experience acute discomfort. The unpleasant sensations we receive act as an early warning system, alerting us to the urgent need to take action to prevent the internal body organs from becoming disastrously chilled or overheated. In addition to encouraging intelligent, voluntary responses, the body also takes certain automatic steps to stabilize its heat level. If the environment becomes too hot, vasodilation occurs. This gives a hotter body surface and encourages heat loss from the skin. Profuse sweating also takes place. We each possess approximately two million sweat glands. Under conditions of intense heat these are capable of secreting a maximum of one litre of sweat per hour. The evaporation of this liquid from the body surface provides another valuable form of heat loss. During the process of acclimatization to a generally hotter environment, we undergo a marked increase in sweating efficiency. This is vitally important because, even in the hottest climates, our internal body temperature can only stand an upward shift of 0·4° Fahrenheit, regardless of our racial origin.

If the environment becomes too cold, we respond with vasoconstriction and with shivering. The vasoconstriction helps to conserve the body heat and the shivering can provide up to three times the resting heat production. If the skin is exposed to the intense cold for any length of time, there is a danger that the prolonged vasoconstriction will lead to frostbite. In the hand region there is an important, built-in, anti-frostbite system. The hands at first respond to intense cold by drastic vasoconstriction; then, after about five minutes, this is reversed and there is strong vasodilation, the hands becoming hot and flushed. (Anyone who has been snow-balling in winter will have experienced this.) The constriction and dilation of the hand region then continues to alternate, the constriction phases curtailing heat loss and the dilation phases preventing frostbite. Individuals living permanently in a cold climate undergo various forms of bodily acclimatiza-tion, including a slightly increased basal metabolic rate.

As our species has spread over the globe, important cultural additions have been made to these biological temperature control mechanisms. The development of fire, clothing, and insulated dwelling-houses has combated heat loss, and ventilation and refrigeration have been used against heat gain. Impressive and dramatic as these advances have been, they have in no way altered our internal body temperature. They have merely served to control the external temperature, so that we can continue to enjoy our primitive primate tempera-ture level in a more diverse range of external conditions.

Before leaving the subject of temperature responses, there is one particular aspect of sweating that should be mentioned. Detailed studies of sweating responses in our species have revealed that they are not as simple as they may first appear. Most areas of the body surface begin to perspire freely under conditions of increased heat, and this is undoubtedly the original, basic response of the sweat-gland system. But certain regions have become reactive to other types of stimu-lation and sweating can occur there regardless of the external temperature. The eating of highly spiced foods, for example, produces its own special pattern of facial sweating. Emotional stress quickly leads to sweating on the palms of the hands, the soles of the feet, the armpits and sometimes also the forehead, but not on other parts of the body. There is a further distinction in the areas of emotional sweating, the palms and the soles differing from the armpits and the forehead. The first two regions respond well *only* to emotional situations, whereas the last two react to both emotional and to temperature stimuli. It is clear from this that the hands and feet have 'borrowed' sweating from the temperature control system and are now using it in a new functional context. The moistening of the palms and soles during stress appears to have become a special feature of the 'ready for anything' response that the body gives when danger threatens. Spitting on the hands before wielding an axe is, in a sense, the non-physiological equivalent of this process.

So sensitive is the palmar sweating response that whole communities or nations may show sudden increases in this reaction if their group security is threatened in some way. During a recent political crisis, when there was a temporary increase in the likelihood of nuclear war, all experiments into palmar sweating at a research institute had to be abandoned because the base level of the response had become so abnormal that the tests would have been meaningless. Having our palms read by a fortune-teller may not tell us much about the future, but having them read by a physiologist can certainly tell us something about our fears for the future.

Chapter Eight

*

ANIMALS

Up to this point we have been considering the naked ape's behaviour towards himself and towards members of his own species—his intra-specific behaviour. It now remains to examine his activities in relation to other animals—his inter-specific behaviour.

All the higher forms of animal life are aware of at least some of the other species with which they share their environment. They regard them in one of five ways: as prey, symbionts, competitors, parasites, or predators. In the case of our own species, these five categories may be lumped together as the 'economic' approach to animals, to which may be added the scientific, aesthetic and symbolic approaches. This wide range of interests has given us an inter-specific involvement unique in the animal world. In order to unravel it and understand it objectively we must tackle it step by step, attitude by attitude.

Because of his exploratory and opportunist nature, the naked ape's list of prey species is immense. At some place, at some time, he has killed and eaten almost any animal you care to mention. From a study of prehistoric remains we know that about half a million years ago, at one site alone, he was hunting and eating species of bison, horse, rhino, deer, bear, sheep, mammoth, camel, ostrich, antelope, buffalo, boar and hyena. It would be pointless to compile a 'species menu' for more recent times, but one feature of our predatory behaviour does deserve mention, namely our tendency to domesticate certain selected prey species. For, although we are likely to eat almost anything palatable on occasion, we have nevertheless limited the bulk of our feeding to a few major animal forms.

Domestication of livestock, involving the organized control and selective breeding of prey, is known to have been practised for at least ten thousand years and, in certain cases, probably much longer. Goats, sheep and reindeer appear to have been the earliest prey species dealt with in this way. Then, with the development of settled agricultural communities, pigs and cattle, including Asiatic buffalo and yak, were added to the list. We have evidence that, in the case of cattle, several distinct breeds had already been developed four thousand years ago. Whereas the goats, sheep and reindeer were transformed directly from hunted prey to herded prey, it is thought that the pigs and cattle began their close association with our species as crop-robbers. As soon as cultivated crops were available, they moved in to take advantage of this rich new food supply, only to be taken over by the early farmers and brought under domestic control themselves.

The only small mammalian prey species to undergo prolonged domestication was the rabbit, but this was apparently a much later development. Amongst the birds, important prey species domesticated thousands of years ago were the chicken, the goose and the duck, with later minor additions of the pheasant, guinea fowl, quail and turkey. The only prey fish with a long history of domestication are the Roman eel, the carp and the goldfish. The latter, however, soon became ornamental rather than gastronomic. The domestication of these fish is limited to the last two thousand years and has played only a small role in the general story of our organized predation.

The second category in our list of inter-specific relation-ships is that of the symbiont. Symbiosis is defined as the association of two different species to their mutual benefit. Many examples of this are known from the animal world, the most famous being the partnership between the tick birds and certain large ungulates such as the rhinoceros, giraffe and buffalo. The birds eat the skin parasites of the ungulates, helping to keep the bigger animals healthy and clean, while the latter provide the birds with a valuable source of food.

Where we ourselves are one of the members of a symbiotic pair, the mutual benefit tends to become biased rather heavily in our favour, but it is nevertheless a separate category, dis-tinct from the more severe prey-predator relationship, since it does not involve the death of the other species concerned. They are exploited, but in exchange for the exploitation we feed and care for them. It is a biased symbiosis because we are in control of the situation and our animal partners usually have little or no choice in the matter.

The most ancient symbiont in our history is undoubtedly the dog. We cannot be certain exactly when our ancestors first began to domesticate this valuable animal, but it appears to be at least ten thousand years ago. The story is a fascinating one. The wild, wolf-like ancestors of the domestic dog must have been serious competitors with our hunting forebears. Both were co-operative pack-hunters of large prey and, at first, little love can have been lost between them. But the wild dogs possessed certain special refinements that our own hunters lacked. They were particularly adept at herding and driving prey during hunting manœuvres and could carry this out at high speed. They also had more delicate senses of smell and hearing. If these attributes could be exploited in exchange for a share in the kill, then the bargain was a good one. Somehow—we do not know exactly how—this came about and an interspecific bond was forged. It is probable that it began as a result of young puppies being brought in to the tribal home base to be fattened as food. The value of these creatures as alert nocturnal watch-dogs would have scored a mark in their favour at an early stage. Those that were allowed to live in a now tamed condition and permitted to accompany the males on their hunting trips would soon show their paces in assisting to track down the prey. Having been hand-reared, the dogs would consider themselves to be members of the naked-ape pack and would co-operate instinctively with their adopted leaders. Selective breeding over a number of generations would soon weed out the trouble-makers and a new, improved stock of increasingly restrained and controllable domestic hunting dogs would arise.

It has been suggested that it was this progression in the dog relationship that made possible the earliest forms of ungulate prey domestication. The goats, sheep and reindeer were under some degree of control before the advent of the true agricultural phase, and the improved dog is envisaged as the vital agent that made this feasible by assisting in the large-scale and long-term herding of these animals. Studies of the driving behaviour of present-day sheepdogs and of wild wolves reveal many similarities in technique and provide strong support for this view.

During more recent times, intensified selective breeding has produced a whole range of symbiotic dog specializations. The primitive all-purpose hunting dog assisted in all stages of the operation, but his later descendants were perfected for one or other of the different components of the overall behaviour sequence. Individual dogs with unusually well-developed abilities in a particular direction were inbred to intensify their special advantages. As we have already seen, those with good qualities in manœuvring became herding dogs, their contribution being confined largely to the round-ing up of domesticated prey (sheepdogs). Others, with a superior sense of smell were inbred as scent-trackers (hounds). Others, with an athletic turn of speed, became coursing dogs and were employed to chase after prey by sight (greyhounds). Another group were bred as prey-spotters, their tendency to 'freeze' when locating the prey being exploited and intensified (setters and pointers). Yet another line was improved as prey-finders and carriers (retrievers). Small breeds were developed as vermin-killers (terriers). The primitive watch-dogs were genetically improved as guard-dogs (mastiffs).

In addition to these widespread forms of exploitation, other dog lines have been selectively bred for more unusual functions. The most extraordinary example is the hairless dog of the ancient New World Indians, a genetically naked breed with an abnormally high skin temperature that was used as a primitive form of hot water-bottle in their sleeping quarters.

In more recent times, the symbiotic dog has earned his keep as beast of burden, pulling sledges or carts, as a messenger or a mine-detector during times of war, as a rescue operator, locating climbers buried under snow, as a police dog, tracking or attacking criminals, as a guide dog, leading the blind, and even as a substitute space-traveller. No other symbiotic species has served us in such a complex and varied way. Even today, with all our technological advances, the dog is still actively employed in most of his functional roles. Many of the hundreds of breeds that can now be distinguished are purely ornamental, but the day of the dog with a serious task to perform is far from over.

So successful has the dog been as a hunting companion that few attempts have been made to domesticate other species in this particular form of symbiosis. The only important exceptions are the cheetah and certain birds of prey, especially the falcon, but in neither case has any progress been made with regard to controlled breeding, let alone selective breeding. Individual training has always been required. In Asia the cormorant, a diving bird, has been used as an active companion in the hunt for fish. Cormorant eggs are taken and hatched out under domestic chickens. The young sea-birds are then hand-reared and trained to catch fish on the end of a line. The fish are brought back to the boats and disgorged, the cormorants having been fitted with a collar to prevent them swallowing their prey. But here again no attempt has been made to improve the stock by selective breeding.

Another ancient form of exploitation involves the use of small carnivores as pest-destroyers. This trend did not gain momentum until the agricultural phase of our history. With the development of large-scale grain storage, rodents became a serious problem and rodent-killers were encouraged. The cat, the ferret and the mongoose were the species that came to our aid and in the first two cases full domestication with selective breeding followed.

Perhaps the most important kind of symbiosis has been the utilization of certain larger species as beasts of burden. Horses, onagers (Asiatic wild asses), donkeys (African wild asses), cattle, including the water buffalo and the yak, reindeer, camels, llamas and elephants have all been subjected to massive exploitation in this way. In most of these cases the original wild types have been 'improved' by careful selective breeding, the exceptions to this rule being the onager and the elephant. The onager was being used as a beast of burden by the ancient Sumerians over four thousand years ago, but was rendered obsolete by the introduction of a more easily controlled species, the horse. The elephant, although still employed as a working animal, has always offered too big a challenge to the stock-breeder and has never been submitted to the pressures of selective breeding.

A further category concerns the domestication of a variety of species as sources of produce. The animals are not killed, so that in this role they cannot be considered as prey. Only certain parts are taken from them: milk from cattle and goats, wool from sheep and alpaca, eggs from chickens and ducks, honey from bees and silk from silk-moths.

In addition to these major categories of hunting companions, pest-destroyers, beasts of burden, and sources of produce, certain animals have entered a symbiotic relationship with our species on a more unusual and specialized basis. The pigeon has been domesticated as a message-carrier. The astonishing homing abilities of this bird have been exploited for thousands of years. This relationship became so valuable in times of war that, during recent epochs, a counter-symbiosis was developed in the form of falcons trained to intercept the message-carriers. In a very different context, Siamese fighting fish and fighting cocks have been selectively bred over a long period as gambling devices. In the realm of medicine, the guinea-pig and the white rat have been widely employed as 'living test-beds' for laboratory experiments.

These, then, are the major symbionts, animals that have been forced into some form of partnership with our ingenious species. The advantage to them is that they cease to be our enemies. Their numbers are dramatically increased. In terms of world populations they are tremendously successful. But it is a qualified success. The price they have paid is their evolutionary freedom. They have lost their genetic in-dependence and, although well fed and cared for, are now subject to our breeding whims and fancies.

The third major category of animal relationships, after prey and symbionts, is that of competitors. Any species which competes with us for food or space, or interferes with the efficient running of our lives, is ruthlessly eliminated. There is no point in listing such species. Virtually any animal that is either inedible or symbiotically useless is attacked and exterminated. This process is continuing today in all parts of the world. In the case of minor competitors, the persecution is haphazard, but serious rivals stand little chance. In the past our closest primate relatives have been our most threatening rivals and it is no accident that today we are the only species surviving in our entire family. Large carnivores have been our other serious competitors and these too have been eliminated wherever the population density of our species has risen above a certain level. Europe, for example, is now virtually denuded of all large forms of carnivores, save for a great seething mass of naked apes.

For the next major category, that of parasites, the future looks even more bleak. Here the fight is intensified and although we may mourn the passing of an attractive food rival, no one will shed a tear over the increasing rarity of the flea. As medical science progresses, the grip of the parasites dwindles. In its wake this brings an added threat to all the other species, for as the parasites go and our health increases, our populations can swell at an even more startling r te, thus accentuating the need to eliminate all the milder competitors.

The fifth major category, the predators, are also on the way out. We have never really constituted a main diet component for any species, and our numbers have never been seriously reduced by predation at any stage in our history, as far as we can tell. But the larger carnivores, such as the big cats and the wild dogs, the bigger members of the crocodile family, the sharks and the more massive birds of prey have nibbled away at us from time to time and their days are clearly numbered. Ironically, the killer that has accounted for more naked-ape deaths than any other (parasites excepted) is one that cannot devour the nutritious corpses it produces. This deadly enemy is the venomous snake and, as we shall see later, this has become the most hated of all higher forms of animal life.

These five categories of inter-specific relationships — prey, symbiont, competitor, parasite and predator — are the ones that can be found to exist between other pairs of species. Basically, we are not unique in these respects. We carry the relationships much further than other species, but they are the same types of relationships. As I said earlier, they can be lumped together as the economic approach to animals. In addition we have our own special approaches, the scientific, the aesthetic and the symbolic.

The scientific and aesthetic attitudes are manifestations of our powerful exploratory drive. Our curiosity, our inquisitiveness, urges us on to investigate all natural phenomena and the animal world has naturally been the focus of much attention in this respect. To the zoologist, all animals are, or should be, equally interesting. To him there are no bad species or good species. He studies them all, exploring them for their own sake. The aesthetic approach involves the same basic exploration, but with different terms of reference. Here, the enormous variety of animal shapes, colours, patterns and movements are studied as objects of beauty rather than as systems for analysis.

The symbolic approach is entirely different. In this case, neither economics nor exploration are involved. The animals are employed instead as personifications of concepts. If a species looks fierce, it becomes a war-symbol. If it looks clumsy and cuddly, it becomes a child-symbol. Whether it is genuinely fierce or genuinely cuddly, matters little. Its true nature is not investigated in this context, for this is not a scientific approach. The cuddly animal may be bristling with razor-sharp teeth and be endowed with a vicious aggressiveness, but providing these attributes are not obvious and its cuddliness is, it is perfectly acceptable as the ideal child-symbol. For the symbolic animal, justice does not have to be done, it has only to appear to be done.

The symbolic attitude to animals was originally christened the 'anthropoidomorphic' approach. Mercifully, this ugly term was later contracted to 'anthropomorphic' which, although still clumsy, is the expression in general use today. It is invariably used in a derogatory sense by scientists who, from their point of view, are fully justified in scorning it. They must retain their objectivity at all costs if they are to make meaningful explorations into the animal world. But this is not as easy as it may sound.

Quite apart from deliberate decisions to use animal forms as idols, images and emblems, there are also subtle, hidden pressures working on us all the time that force us to see other species as caricatures of ourselves. Even the most sophisticated scientist is liable to say, 'Hallo, old boy' when greeting his dog. Although he knows perfectly well that the animal cannot understand his words, he cannot resist the temptation. What is the nature of these anthropomorphic pressures and why are they so difficult to overcome? Why do some creatures make us say 'Aah' and others make us say 'Ugh!'? This is no trivial consideration. A vast amount of our present culture's inter-specific energies are involved here. We are passionate animal lovers and animal haters, and these involvements cannot be explained on the basis of economic and exploratory considerations alone. Clearly some kind of unsuspected, basic response is being triggered off inside us by the specific signals we are receiving. We delude ourselves that we are responding to the animal as an animal. We declare that it is charming, irresistible, or horrible, but what makes it so?

In order to find the answer to this question we must first assemble some facts. What exactly are the animal loves and animal hates of our culture and how do they vary with age and sex? Quantitative evidence is required on a large scale if reliable statements are to be made on this topic. To obtain such evidence an investigation was carried out involving 80,000 British children between the ages of four and fourteen. During a zoo television programme they were asked the simple questions: 'Which animal do you like most?' and 'Which animal do you dislike most?' From the massive response to this inquiry a sample of 12,000 replies to each question was selected at random and analysed.

Dealing first with the interspecific 'loves', how did the various groups of animals fare? The figures are as follows: 97·15 per cent of all the children quoted a mammal of some kind as their top favourite. Birds accounted for only 1·6 per cent, reptiles 1·0 per cent, fish 0·1 per cent, invertebrates 0·1 per cent, and amphibians 0·05 per cent. Obviously there is something special about mammals in this context.

(It should perhaps be pointed out that the replies to the questions were written, not spoken, and it was sometimes difficult to identify the animals from the names given, especially in the case of very young children. It was easy enough to decipher loins, hores, bores, penny kings, panders, tapers and leapolds, but almost impossible to be certain of the species referred to as bettle twigs, the skipping worm, the otamus, or the coco-cola beast. Entries supporting these appealing creatures were reluctantly rejected.)

If we now narrow our sights to the 'top ten animal loves' the figures emerge as follows: 1. Chimpanzee (13·5 per cent). 2. Monkey (13 per cent). 3. Horse (9 per cent). 4. Bushbaby (8 per cent). 5. Panda (7·5 per cent). 6. Bear (7 per cent).

7. Elephant (6 per cent). 8. Lion (5 per cent). 9. Dog (4 per cent). 10. Giraffe (2·5 per cent).

It is immediately clear that these preferences do not reflect powerful economic or aesthetic influences. A list of the ten most important economic species would read very differently. Nor are these animal favourites the most elegant and brightly coloured of species. They include instead a high proportion of rather clumsy, heavy-set and dully coloured forms. They are, however, well endowed with anthropomorphic features and it is to these that the children are responding when making their choices. This is not a conscious process. Each of the species listed provides certain key stimuli strongly reminiscent of special properties of our own species, and to these we react automatically without any realization of what it is exactly that appeals to us. The most significant of these anthropomorphic features in the top ten animals are as follows:

1. They all have hair, rather than feathers or scales. 2. They have rounded outlines (chimpanzee, monkey, bushbaby, panda, bear, elephant). 3. They have flat faces (chimpanzee, monkey, bushbaby, bear, panda, lion). 4. They have facial expressions (chimpanzee, monkey, horse, lion, dog). 5. They can 'manipulate' small objects (chimpanzee, monkey, bushbaby, panda, elephant). 6. Their postures are in some ways, or at some times, rather vertical (chimpanzee, monkey, bushbaby, panda, bear, giraffe).

The more of these points a species can score, the higher up the top ten list it comes. Non-mammalian species fare badly because they are weak in these respects. Amongst the birds, the top favourites are the penguin (0·8 per cent) and the parrot (0·2 per cent). The penguin achieves the number one avian position because it is the most vertical of all the birds. The parrot also sits more vertically on its perch than most birds and it has several other special advantages. Its beak shape gives it an unusually flattened face for a bird. It also feeds in a strange way, bringing its foot up to its mouth rather than lowering its head, and it can mimic our vocalizations. Unfortunately for its popularity, it lowers itself into a more horizontal posture when walking and in this way loses points heavily to the vertically waddling penguin.

Amongst the top mammals there are several special points worth noting. Why, for instance, is the lion the only one of the big cats to be included? The answer appears to be that it alone, in the male, has a heavy mane of hair surrounding the head region. This has the effect of flattening the face (as is clear from the way lions are portrayed in children's drawings) and helps to score extra points for this species.

Facial expressions are particularly important, as we have already seen in earlier chapters, as basic forms of visual communication in our species. They have evolved in a complex form in only a few groups of mammals — the higher primates, the horses, the dogs and the cats. It is no accident that five of the top ten favourites belong to these groups. Changes in facial expression indicate changes in mood and this provides a valuable link between the animal and ourselves, even though the correct significance of the expressions is not always precisely understood.

As regards manipulative ability, the panda and the elephant are unique cases. The former has evolved an elongated wrist bone with which it can grasp the thin bamboo sticks on which it feeds. A structure of this kind is found nowhere else in the animal kingdom. It gives the flat-footed panda the ability to hold small objects and bring them up to its mouth while sitting in a vertical posture. Anthropomorphically this scores heavily in its favour. The elephant is also capable of 'manipulating' small objects with its trunk, another unique structure, and taking them up to its mouth.

The vertical posture so characteristic of our species gives any other animal that can adopt this position an immediate anthropomorphic advantage. The primates in the top ten list, the bears and the panda all sit up vertically on frequent occasions. Sometimes they may even stand vertically or go so far as to take a few faltering steps in this position, all of which helps them to score valuable points. The giraffe, by virtue of its unique body proportions, is, in a sense, permanently vertical. The dog, which achieves such a high anthropomorphic score for its social behaviour, has always been something of a postural disappointment. It is uncompromisingly horizontal. Refusing to accept defeat on this point, our ingenuity went to work and soon solved the problem — we taught the dog to sit up and beg. In our urge to anthropomorphize the poor creature, we went further still. Being tailless ourselves, we started docking its tail. Being flat-faced ourselves, we employed selective breeding to reduce the bone structure in the snout region. As a result, many dog breeds are now abnormally flat-faced. Our anthropomorphic desires are so demanding that they have to be satisfied, even at the expense of the animals' dental efficiency. But then we must recall that this approach to animals is a purely selfish one. We are not seeing animals as animals, but merely as reflections of ourselves, and if the mirror distorts too badly we either bend it into shape or discard it.

So far we have been considering the animal loves of children of all ages between four and fourteen. If we now split up the responses to these favourite animals, separating them into age groups, some remarkably consistent trends emerge. For certain of the animals there is a steady decrease in preference with the increasing age of the children. For others there is a steady rise.

The unexpected discovery here is that these trends show a

marked relationship with one particular feature of the preferred animals, namely their body size. The younger children
prefer the bigger animals and the older children prefer the
smaller ones. To illustrate this we can take the figures for the
two largest of the top ten forms, the elephant and the giraffe,
and two of the smallest, the bushbaby and the dog. The
elephant, with an overall average rating of 6 per cent, starts
out at 15 per cent with the four-year-olds and then falls
smoothly to 3 per cent with the fourteen-year-olds. The
giraffe shows a similar drop in popularity from 10 per cent to 1
per cent. The bushbaby, on the other hand, starts at only 4·5
per cent with the four-year-olds and then rises gradually to 11
per cent with the fourteen-year-olds. The dog rises from
0·5 to 6·5 per cent. The medium-sized animals amongst the
top ten favourites do not show these marked trends.

We can sum up the findings so far by formulating two
principles. The first law of animal appeal states that 'The
popularity of an animal is directly correlated with the number
of anthropomorphic features it possesses.' The second law of
animal appeal states that 'The age of a child is inversely
correlated with the size of the animal it most prefers.'

How can we explain the second law? Remembering that
the preference is based on a symbolic equation, the simplest
explanation is that the smaller children are viewing the animals
as parent-substitutes and the older children are looking upon
them as child-substitutes. It is not enough that the animal must
remind us of our own species, it must remind us of a special
category within it. When the child is very young, its parents
are all-important protective figures. They dominate the
child's awareness. They are large, friendly animals, and large
friendly animals are therefore easily identified with parental
figures. As the child grows it starts to assert itself, to compete
with its parents. It sees itself in control of the situation, but it is

difficult to control an elephant or a giraffe. The preferred animal has to shrink down to a manageable size. The child, in a strangely precocious way, becomes the parent itself. The animal has become the symbol of *its* child. The real child is too young to be a real parent, so instead it becomes a symbolic parent. Ownership of the animal becomes important and pet-keeping develops as a form of 'infantile parentalism'. It is no accident that, since becoming available as an exotic pet, the animal previously known as the galago has now acquired the popular name of bush*baby*. (Parents should be warned from this that the pet-keeping urge does not arrive until late in childhood. It is a grave error to provide pets for very young children, who respond to them as objects for destructive exploration, or as pests.)

There is one striking exception to the second law of animal appeal and that concerns the horse. The response to this animal is unusual in two ways. When analysed against increase in age of children, it shows a smooth rise in popularity followed by an equally smooth fall. The peak coincides with the onset of puberty. When analysed against the different sexes, it emerges that it is three times as popular with girls as with boys. No other animal love shows anything approaching this sex difference. Clearly there is something unusual about the response to the horse and it requires separate consideration.

The unique feature of the horse in the present context is that it is something to be mounted and ridden. This applies to none of the other top ten animals. If we couple this observation with the facts that its popularity peak coincides with puberty and that there is a strong sexual difference in its appeal, we are forced to the conclusion that the response to the horse must involve a strong sexual element. If a symbolic equation is being made between mounting a horse and sexual mounting, then it is perhaps surprising that the animal has a greater appeal for girls. But the horse is a powerful, muscular and dominant animal and is therefore more suited to the male role. Viewed objectively, the act of horse-riding consists of a long series of rhythmic movements with the legs wide apart and in close contact with the body of the animal. Its appeal for girls appears to result from the combination of its masculinity and the nature of the posture and actions performed on its back. (It must be stressed here that we are dealing with the child population as a whole. One child in every eleven preferred the horse to all other animals. Only a small fraction of this percentage would ever actually own a pony or a horse. Those that do, quickly learn the many more varied rewards that go with this activity. If, as a result, they become addicted to horse-riding, this is not, of course, necessarily significant in the context we have been discussing.)

It remains to explain the fall in popularity of the horse following puberty. With increasing sexual development, it might be expected to show further increases in popularity, rather than a decrease. The answer can be found by comparing the graph for horse love with the curve for sex play in children. They match one another remarkably well. It would seem that, with the growth of sexual awareness, and the characteristic sense of privacy that comes to surround teenage sexual feelings, the response to the horse declines along with the decline in overt sex-play 'romping'. It is significant here that the appeal of monkeys also suffers a decline at this point. Many monkeys have particularly obtrusive sexual organs, including large, pink, sexual swellings. For the younger child these have no significance and the monkeys' other powerful anthropomorphic features can operate unhindered. But for older children the conspicuous genitals become a source of embarrassment and the popularity of these animals suffers as a consequence.

This, then, is the situation with regard to animal 'loves' in children. For adults, the responses become more varied and sophisticated, but the basic anthropomorphism persists. Serious naturalists and zoologists bewail this fact, but providing it is fully realized that symbolic responses of this kind tell us nothing about the true nature of the different animals concerned, they do little harm and provide a valuable subsidiary outlet for emotional feelings.

Before considering the other side of the coin – the animal 'hates' – there is one criticism that must be answered. It could be argued that the results discussed above are of purely cultural significance, and have no meaning for our species as a whole. As regards the exact identity of the animals involved this is true. To respond to a panda, it is obviously necessary to learn of its existence. There is no inborn panda response. But this is not the point. The choice of the panda may be culturally determined, but the *reasons* for choosing it do reflect a deeper, more biological process at work. If the investigation were repeated in another culture, the favourite species might be different, but they would still be selected according to our fundamental symbolic needs. The first and second law of animal appeal would still operate.

Turning now to animal 'hates', we can subject the figures to a similar analysis. The top ten most disliked animals are as follows: 1. Snake (27 per cent). 2. Spider (9·5 per cent). 3. Crocodile (4·5 per cent). 4. Lion (4·5 per cent). 5. Rat (4 per cent). 6. Skunk (3 per cent). 7. Gorilla (3 per cent). 8. Rhinoceros (3 per cent). 9. Hippopotamus (2·5 per cent). 10. Tiger (2·5 per cent).

These animals share one important feature: they are dangerous. The crocodile, the lion and the tiger are carnivorous killers. The gorilla, the rhinoceros and the hippopotamus can easily kill if provoked. The skunk indulges in a violent form of chemical warfare. The rat is a pest that spreads disease. There are venomous snakes and poisonous spiders.

Most of these creatures are also markedly lacking in the anthropomorphic features that typify the top ten favourites. The lion and the gorilla are exceptions. The lion is the only form to appear in both the top ten lists. The ambivalence of the response to this species is due to this animal's unique combination of attractive anthropomorphic characters and violent predatory behaviour. The gorilla is strongly endowed with anthropomorphic characters, but unfortunately for him his facial structure is such that he appears to be in a constantly aggressive and fearsome mood. This is merely an accidental outcome of his bone structure and bears no relationship to his true (and rather gentle) personality, but combined with his great physical strength it immediately converts him into a perfect symbol of savage brute force.

The most striking feature of the list of top ten hates is the massive response to the snake and the spider. This cannot be explained solely on the basis of the existence of dangerous species. Other forces are at work. An analysis of the reasons given for hating these forms reveals that snakes are disliked because they are 'slimy and dirty' and spiders are repulsive because they are 'hairy and creepy'. This must mean either that they have a strong symbolic significance of some kind, or alternatively that we have a powerful inborn response to avoid these animals.

The snake has long been thought of as a phallic symbol. Being a poisonous phallus, it has represented unwelcome sex, which may be a partial explanation for its unpopularity; but there is more to it than this. If we examine the different levels of snake hatred in children between the ages of four and fourteen, it emerges that the peak of unpopularity comes early, long before puberty is reached. Even at four, the hate level is high—around 30 per cent—and it then climbs slightly, reaching its peak at age six. From then on it shows a smooth decline, sinking to well below 20 per cent by the age of fourteen. There is little difference between the sexes, although at each age level the response from girls is slightly stronger than the response from boys. The arrival of puberty appears to have no impact on the response in either sex.

From this evidence it is difficult to accept the snake simply as a strong sexual symbol. It seems more likely that we are dealing here with an inborn aversion response of our species towards snake-like forms. This would explain not only the early maturation of the reaction, but also the enormously high level of the response when compared with all other animal hates and loves. It would also fit with what we know of our closest living relatives, the chimpanzees, gorillas and orangutans. These animals also exhibit a great fear of snakes and here again it matures early. It is not seen in the very young apes, but is fully developed by the time they are a few years old and have reached the stage where they are beginning to make brief sorties away from the security of their mothers' bodies. For them an aversion response clearly has an important survival value and would also have been a great benefit to our early ancestors. Despite this, it has been argued that the snake reaction is not inborn, but merely a cultural phenomenon resulting from individual learning. Young chimpanzees reared under abnormally isolated conditions have reputedly failed to show the fear response when first exposed to snakes. But these experiments are not very convincing. In some instances, the chimpanzees have been too young when first tested. Had they been re-tested a few years later, the reaction may well have been present. Alternatively, the effects of isolation may have been so severe that the young animals in question were virtually mental defectives. Such experiments are based on a fundamental misconception about the nature of inborn responses, which do not mature in an encapsulated form, irrespective of the outside environment. They should be thought of more as inborn susceptibilities. In the case of the snake response, it may be necessary for the young chimpanzee, or child, to encounter a number of different frightening objects in its early life and to learn to respond negatively to these. The inborn element in the snake case would then manifest itself in the form of a much more massive response to this stimulus than to others. The snake fear would be out of all proportion to the other fears, and this disproportionateness would be the inborn factor. The terror produced in normal young chimpanzees by exposure to a snake and the intense hatred of snakes exhibited by our own species is difficult to explain in any other way.

The reaction of children to spiders takes a rather different course. Here there is a marked sex difference. In boys there is an increase in spider hatred from age four to fourteen, but it is slight. The level of the reaction is the same for girls up to the age of puberty, but it then shows a dramatic rise, so that by the age of fourteen it is double that of the boys. Here we do seem to be dealing with an important symbolic factor. In evolutionary terms, poisonous spiders are just as dangerous to males as to females. There may or may not be an inborn response to these creatures in both sexes, but it cannot explain the spectacular leap in spider hatred that accompanies female puberty. The only clue here is the repeated female reference to spiders being nasty, hairy things. Puberty is, of course, the stage when tufts of body hair are beginning to sprout on both boys and girls. To children, body hairiness must appear as an essentially masculine character. The growth of hair on the body of a young girl would therefore have a more disturbing (unconscious) significance for her than it would in the case of a boy. The long legs of a spider are more hairlike and more obvious than those of other small creatures such as flies, and it would as a result be the ideal symbol in this role.

These, then, are the loves and the hatreds we experience when encountering or contemplating other species. Combined with our economic, scientific and aesthetic interests, they add up to a uniquely complex inter-specific involvement, and one which changes as we grow older. We can sum this up by saying that there are 'seven ages' of inter-specific reactivity. The first age is the *infantile phase*, when we are completely dependent on our parents and react strongly to very big animals, employing them as parent symbols. The second is the *infantile-parental phase*, when we are beginning to compete with our parents and react strongly to small animals that we can use as child-substitutes. This is the age of pet-keeping. The third age is the *objective pre-adult phase*, the stage where the exploratory interests, both scientific and aesthetic, come to dominate the symbolic. It is the time for bug-hunting, microscopes, butterfly-collecting and aquaria. The fourth is the *young adult phase*. At this point the most important animals are members of the opposite sex of our own species. Other species lose ground here, except in a purely commercial or economic context. The fifth is the *adult parental phase*. Here symbolic animals enter our lives again, but this time as pets for our children. The sixth age is the *post-parental phase*, when we lose our children and may turn once more to animals as child-substitutes to replace them. (In the case of childless adults, the use of animals as child-substitutes may, of course, begin earlier.) Finally, we come to the seventh age, the *senile phase*, which is characterized by a heightened interest in animal preservation and conservation. At this point the interest is focused on those species which are in danger of extermination. It makes little difference whether, from other points of view, they are attractive or repulsive, useful or useless, providing their numbers are few and becoming fewer. The increasingly rare rhinoceros and gorilla, for example, that are so disliked by children, become the centre of attention at this stage. They have to be 'saved'. The symbolic equation involved here is obvious enough: the senile individual is about to become personally extinct and so employs rare animals as symbols of his own impending doom. His emotional concern to save them from extinction reflects his desire to extend his own survival.

During recent years, interest in animal conservation has spread to some extent into the lower age groups, apparently as a result of the development of immensely powerful nuclear weapons. Their huge destructive potential threatens all of us, regardless of age, with the possibility of immediate extermination, so that now we all have an emotional need for animals that can serve as rarity symbols.

This observation should not be interpreted as implying that this is the only reason for the conservation of wild life. There are, in addition, perfectly valid scientific and aesthetic reasons why we should wish to give aid to unsuccessful species. If we are to continue to enjoy the rich complexities of the animal world and to use wild animals as objects of scientific and aesthetic exploration, we must give them a helping hand. If we allow them to vanish, we shall have simplified our environment in a most unfortunate way. Being an intensely investigatory species, we can ill afford to lose such a valuable source of material.

Economic factors are also sometimes mentioned when conservation problems are under discussion. It is pointed out that intelligent protection and controlled cropping of wild species can assist the protein-starved populations in certain parts of the world. While this is perfectly true on a short-term basis, the long-term picture is more gloomy. If our numbers continue to increase at the present frightening rate, it will eventually become a matter of choosing between us and them. No matter how valuable they are to us symbolically, scientifically or aesthetically, the economics of the situation will shift against them. The blunt fact is that when our own species density reaches a certain pitch, there will be no space left for other animals. The argument that they constitute an essential source of food does not, unhappily, stand up to close scrutiny. It is more efficient to eat plant food direct, than to convert it into animal flesh and then eat the animals. As the demand for living space increases still further, even more drastic steps will ultimately have to be taken and we shall be driven to synthesizing our foodstuffs. Unless we can colonize other planets on a massive scale and spread the load, or seriously check our population increase in some way, we shall, in the not-too-far distant future, have to remove all other forms of life from the earth.

If this sounds rather melodramatic, consider the figures involved. At the end of the seventeenth century the world population of naked apes was only 500 million. It has now risen to 3,000 million. Every twenty-four hours it increases by another 150,000. (The inter-planetary emigration authorities would find this figure a daunting challenge.) In 260 years' time, if the rate of increase stays steady — which is unlikely — there will be a seething mass of 400,000 million naked apes crowding the face of the earth. This gives a figure of 11,000 individuals to every square mile of the entire land surface. To put it another way, the densities we now experience in our major cities would exist in every corner of the globe. The consequence of this for all forms of wild life is obvious. The effect it would have on our own species is equally depressing.

We need not dwell on this nightmare: the possibility of its becoming a reality is remote. As I have stressed throughout this book, we are, despite all our great technological advances, still very much a simple biological phenomenon. Despite our grandiose ideas and our lofty self-conceits, we are still humble animals, subject to all the basic laws of animal behaviour. Long before our populations reach the levels envisaged above we shall have broken so many of the rules that govern our biological nature that we shall have collapsed as a dominant species. We tend to suffer from a strange complacency that this can never happen, that there is something special about us, that we are somehow above biological control. But we are not. Many exciting species have become extinct in the past and we are no exception. Sooner or later we shall go, and make way for something else. If it is to be later rather than sooner, then we must take a long, hard look at ourselves as biological specimens and gain some understanding of our limitations. This is why I have written this book, and why I have deliberately insulted us by referring to us as naked apes, rather than by the more usual name we use for ourselves. It helps to keep a sense of proportion and to force us to consider what is going on just below the surface of our lives. In my enthusiasm I may, perhaps, have overstated my case. There are many praises I could have sung, many magnificent achievements I could have described. By omitting them I have inevitably given a one-sided picture. We are an extraordinary species and I do not wish to deny it, or to belittle us. But these things have been said so often. When the coin is tossed it always seems to come up heads, and I have felt that it was high time we turned it over and looked at the other side. Unfortunately, because we are so powerful and so successful when compared with other animals, we find the contemplation of our humble origins somehow offensive, so that I do not expect to be thanked for what I have done. Our climb to the top has been a get-rich-quick story, and, like all *nouveaux riches*, we are very sensitive about our background. We are also in constant danger of betraying it.

Optimism is expressed by some who feel that since we have evolved a high level of intelligence and a strong inventive urge, we shall be able to twist any situation to our advantage; that we are so flexible that we can re-mould our way of life to fit any of the new demands made by our rapidly rising species-status; that when the time comes, we shall manage to cope with the over-crowding, the stress, the loss of our privacy and independence of action; that we shall re-model our behaviour patterns and live like giant ants; that we shall control our aggressive and territorial feelings, our sexual impulses and our parental tendencies; that if we have to become battery-chicken apes, we can do it; that our intelligence can dominate all our basic biological urges. I submit that this is rubbish. Our raw animal nature will never permit it. Of course, we are flexible. Of course, we are behavioural opportunists, but there are severe limits to the form our opportunism can take. By stressing our biological features in this book, I have tried to show the nature of these restrictions. By recognizing them clearly and submitting to them, we shall stand a much better chance of survival. This does not imply a naive 'return to nature'. It simply means that we should tailor our intelligent opportunist advances to our basic behavioural requirements. We must somehow improve in quality rather than in sheer quantity. If we do this, we can continue to progress technologically in a dramatic and exciting way without denying our evolutionary inheritance. If we do not, then our suppressed biological urges will build up and up until the dam bursts and the whole of our elaborate existence is swept away in the flood.

Appendix: Literature

*

It is impossible to list all the many works that have been of assistance in writing *The Naked Ape*, but some of the more important ones are arranged below on a chapter-by-chapter and topic-by-topic basis. Detailed references for these publications are given in the bibliography that follows this appendix.

Chapter One. Origins

Classification of primates: Morris, 1965. Napier and Napier, 1967.

Evolution of primates: Dart and Craig, 1959. Eimerl and DeVore, 1965. Hooton, 1947. Le Gros Clark, 1959. Morris and Morris, 1966. Napier and Napier, 1967. Oakley, 1961. Read, 1925. Washburn, 1962 and 1964. Tax, 1960.

Carnivore behaviour: Guggisberg, 1961. Kleiman, 1966. Kruuk, 1966. Leyhausen, 1956. Lorenz, 1954. Moulton, Ashton and Eayrs, 1960. Neuhaus, 1953. Young and Goldman, 1944.

Primate behaviour: Morris, 1967. Morris and Morris, 1966. Schaller, 1963. Southwick, 1963. Yerkes and Yerkes, 1929. Zuckerman, 1932.

Chapter Two. Sex

Animal courtship: Morris, 1956.
Sexual responses: Masters and Johnson, 1966.
Sexual pattern frequencies: Kinsey et al., 1948 and 1953.
Self-mimicry: Wickler, 1963 and 1967.
Mating postures: Ford and Beach, 1952.
Odour preferences: Monicreff, 1965.
Chastity devices: Gould and Pyle, 1896.
Homosexuality: Morris, 1955.

Chapter Three. Rearing

Suckling: Gunther, 1955. Lipsitt, 1966.
Heart-beat response: Salk, 1966.
Growth rates: Harrison, Weiner, Tanner and Barnicott, 1964.
Sleep: Kleitman, 1963.
Stages of development: Shirley, 1933.
Development of vocabulary: Smith, 1926.
Chimpanzee vocal imitations: Hayes, 1952.
Crying, smiling and laughing: Ambrose, 1960.
Facial expressions in primates: van Hooff, 1962.
Group density in children: Hutt and Vaizey, 1966.

Chapter Four. Exploration

Neophilia and neophobia: Morris, 1964.
Ape picture-making: Morris, 1962.
Infant picture-making: Kellogg, 1955.
Chimpanzee exploratory behaviour: Morris and Morris, 1966.
Isolation during infancy: Harlow, 1958.
Stereotyped behaviour: Morris, 1964 and 1966.

Chapter Five. Aggression

Primate aggression: Morris and Morris, 1966.
Autonomic changes: Cannon, 1929.
Origin of signals: Morris, 1956 and 1957.
Displacement activities: Tinbergen, 1951.
Facial expressions: van Hooff, 1962.
Eye-spot signals: Coss, 1965.
Reddening of buttocks: Comfort, 1966.
Redirection of aggression: Bastock, Morris and Moynihan, 1953.
Over-crowding in animals: Calhoun, 1962.

Chapter Six. Feeding

Male association patterns: Tiger, 1967.
Organs of taste and smell: Wyburn, Pickford and Hirst, 1964.
Cereal diets: Harrison, Weiner, Tanner and Barnicott, 1964.

Chapter Seven. Comfort

Social grooming: van Hooff, 1962. Sparks, 1963. (I am particularly indebted to Jan van Hooff for inventing the term 'Grooming talk'.)
Skin glands: Montagna, 1956.
Temperature responses: Harrison, Weiner, Tanner and Barnicott, 1964.
'Medical' aid in chimpanzees: Miles, 1963.

Chapter Eight. Animals

Domestication: Zeuner, 1963.
Animal likes: Morris and Morris, 1966.
Animal dislikes: Morris and Morris, 1965.
Animal phobias; Marks, 1966.
Population explosion: Fremlin, 1965.

Bibliography

*

AMBROSE, J. A., 'The smiling response in early human infancy' (Ph.D.thesis, London University, 1960), pp. 1–660.

BASTOCK, M., D. MORRIS, and M. MOYNIHAN, 'Some comments on conflict and thwarting in animals'. *Behaviour* 6 (1953), pp. 66–84.

BEACH, F. A. (editor), *Sex and Behaviour* (Wiley, New York, 1965).

BERELSON, B. and G. A. STEINER, *Human Behaviour* (Harcourt, Brace and World, New York, 1964).

CALHOUN, J. B., 'A "behavioral sink",' in *Roots of Behaviour*, (ed. E. L. Bliss) (Harper and Brothers, New York, 1962), pp. 295–315.

CANNON, W. B., *Bodily Changes in Pain, Hunger, Fear and Rage* (Appleton-Century, New York, 1929).

CLARK, W. E. LE GROS. (1959). *The Antecedents of Man*. Edinburgh University Press. Quadrangle, Chicago (1960).

COLBERT, E. H., *Evolution of the Vertebrates* (Wiley, New York, 1955).

COMFORT, A., *Nature and Human Nature* (Weidenfeld and Nicolson, 1966).

COSS, R. G., *Mood Provoking Visual Stimuli* (University of California, 1965).

DART, R. A. and D. CRAIG, *Adventures with the Missing Link* (Hamish Hamilton, 1959).

EIMERL, S. and I. DEVORE, *The Primates* (Time Life, New York, 1965).

FORD, C. S., and F. A. BEACH. (1952). *Patterns of Sexual Behaviour.* Eyre and Spottiswoode, London. Harper & Row, New York (1951).

FREMLIN, J. H., 'How many people can the world support?' *New Scientist* 24 (1965), pp. 285–7.

GOULD, G. M. and W. L. PYLE, *Anomalies and Curiosities of Medicine* (Saunders, Philadelphia, 1896).

GUGGISBERG, C. A. W. (1961). *Simba: The Life of the Lion.* Bailey Bros. and Swinfen, London. Chilton, Philadelphia (1963).

GUNTHER, M., 'Instinct and the nursing couple'. *Lancet* (1955), pp. 575–8.

HARDY, A. C., 'Was man more aquatic in the past?' *New Scientist* 7 (1960), pp. 642–5.

HARLOW, H. F., 'The nature of love'. *Amer. Psychol.* 13 (1958), pp. 673–85.

HARRISON, G. A., J. S. WEINER, J. M. TANNER and N. A. BARNICOTT, *Human Biology* (Oxford University Press, 1964).

HAYES, C., *The Ape in our House* (Gollancz, 1952).

HOOTON, E. A., *Up from the Ape* (Macmillan, New York, 1947).

HOWELLS, W., *Mankind in the Making* (Secker and Warburg, 1960).

HUTT, C. and M. J. VAIZEY, 'Differential effects of group density on social behaviour'. *Nature* 209 (1966), pp. 1371–2.

KELLOGG, R., *What Children Scribble and Why* (Author's edition, San Francisco, 1955).

KINSEY, A. C., W. B. POMEROY and C. E. MARTIN, *Sexual Behaviour in the Human Male* (Saunders, Philadelphia, 1948).

KINSEY, A. C., W. B. POMEROY, C. E. MARTIN and P. H. GEBHARD, *Sexual Behaviour in the Human Female* (Saunders, Philadelphia, 1953).

KLEIMAN, D., 'Scent marking in the Canidae'. *Symp. Zool. Soc.* 18 (1966), pp. 167–77.

KLEITMAN, N., *Sleep and Wakefulness* (Chicago University Press, 1963).

KRUUK, H., 'Clan-system and feeding habits of Spotted Hyenas'. *Nature* 209 (1966), pp. 1257–8.

LEYHAUSEN, P., *Verhaltensstudien an Katzen* (Paul Parey, Berlin, 1956).

LIPSITT, L., 'Learning processes of human newborns'. *Merril-Palmer Quart. Behav. Devel.* 12 (1966), pp. 45–71.

LORENZ, K. (1952). *King Solomon's Ring*. Methuen, London. Crowell, New York (1952).

LORENZ, K., *Man Meets Dog* (Methuen, 1954).

MARKS, I. M. and M. G. GELDER, 'Different onset ages in varieties of phobias'. *Amer. J. Psychiat.* (July 1966).

MASTERS, W. H., and V. E. Johnson. (1966). *Human Sexual Response*. Churchill, London. Little Brown, Boston (1966).

MILES, W. R., 'Chimpanzee behaviour: removal of foreign body from companion's eye'. *Proc. Nat. Acad. Sci.* 49 (1963), pp. 840–3.

MONICREFF, R. W., 'Changes in olfactory preferences with age'. *Rev. Laryngol.* (1965), pp. 895–904.

MONTAGNA, W. (1956). *The Structure and Function of Skin*. Academic Press, London. Academic Press, New York (1956).

MONTAGU, M. F. A., *An Introduction to Physical Anthropology* (Thomas, Springfield, 1945).

MORRIS, D., 'The causation of pseudofemale and pseudomale behaviour'. *Behaviour* 8 (1955), pp. 46–56.

MORRIS, D., 'The function and causation of courtship ceremonies'. Fondation Singer Polignac Colloque Internat. sur L'Instinct, June 1954 (1956), pp. 261–86.

MORRIS, D., 'The feather postures of birds and the problem of the origin of social signals'. *Behaviour* 9 (1956), pp. 75–113.

MORRIS, D., ' "Typical Intensity" and its relation to the problem of ritualization'. *Behaviour* 11 (1957), pp. 1–12.

MORRIS, D. (1962). *The Biology of Art*. Methuen, London. Knopf, New York (1962).

MORRIS, D., 'The response of animals to a restricted environment'. *Symp. Zool. Soc. Lond.* 13 (1964), pp. 99–118.

MORRIS, D. (1965). *The Mammals: A Guide to the Living Species*. Hodder and Stoughton, London. Harper & Row, New York (1965).

MORRIS, D., 'The rigidification of behaviour'. *Phil. Trans. Roy. Soc. London*, B. 251 (1966), pp. 327–30.

MORRIS, D. (editor), *Primate Ethology* (Weidenfeld and Nicolson, 1967).

MORRIS, R. and D. MORRIS. (1965). *Men and Snakes*. Hutchinson, London. McGraw-Hill, New York (1965).

MORRIS, R. and D. MORRIS. (1966). *Men and Apes*. Hutchinson, London. McGraw-Hill, New York (1966).

MORRIS, R. and D. MORRIS, *Men and Pandas* (Hutchinson, 1966).

MOULTON, D. G., E. H. ASHTON and J. T. EAYRS, 'Studies in olfactory acuity. 4. Relative detectability of n-Aliphatic acids by dogs'. *Anim. Behav.* 8 (1960), pp. 117–28.

NAPIER, J. and P. NAPIER, *Primate Biology* (Academic Press, 1967).

NEUHAUS, W., 'Über die Riechschärfe der Hunden für Fettsäuren'. *Z. vergl. Physiol.* 35 (1953), pp. 527–52.

OAKLEY, K. P., *Man the Toolmaker*. Brit. Mus. (Nat. Hist.), 1961.

READ, C., *The Origin of Man* (Cambridge University Press, 1925).

ROMER, A. S., *The Vertebrate Story* (Chicago University Press, 1958).

RUSSELL, C., and W. M. S. RUSSELL. (1961). *Human Behaviour*. André Deutsch, London. Little Brown, Boston (1961).

SALK, L., 'Thoughts on the concept of imprinting and its place in early human development'. *Canad. Psychiat. Assoc. J.* 11 (1966), pp. 295–305.

SCHALLER, G., *The Mountain Gorilla* (Chicago University Press, 1963).

SHIRLEY, M. M., 'The first two years, a study of twenty-five babies'. Vol. 2, *Intellectual development. Inst. Child Welf. Mongr.*, Serial No. 8 (University of Minnesota Press, Minneapolis, 1933).

SMITH, M. E., 'An investigation of the development of the sentence and the extent of the vocabulary in young children'. *Univ. Iowa Stud. Child. Welf.* 3, No. 5 (1926).

SPARKS, J., 'Social grooming in animals'. *New Scientist* 19 (1963), pp. 235–7.

SOUTHWICK, C. H. (editor), *Primate Social Behaviour* (van Nostrand, Princeton, 1963).

TAX, S. (editor), *The Evolution of Man* (Chicago University Press, 1960).

TIGER, L., Research report: Patterns of male association. *Current Anthropology* (vol. VIII, No. 3, June 1967).

TINBERGEN, N., *The Study of Instinct* (Oxford University Press, 1951).

VAN HOOFF, J., 'Facial expressions in higher primates'. *Symp. Zool. Soc. Lond.* 8 (1962), pp. 97–125.

WASHBURN, S. L. (editor) (1962). *Social Life of Early Man*. Methuen, London. Aldine Publishing Co., Chicago (1962).

WASHBURN, S. L. (editor) (1964). *Classification and Human Evolution*. Methuen, London. Aldine Publishing Co., Chicago (1964).

WICKLER, W., 'Die biologische Bedeutung auffallend far-
biger, nackter Hautstellen und innerartliche Mimikry der
Primaten'. *Die Naturwissenschaften* 50 (13) (1963), pp.
481–2.

WICKLER, W., Socio-sexual signals and their intra-specific
imitation among primates. In *Primate Ethology*, (ed. D.
Morris) (Weidenfeld & Nicolson, 1967), pp. 68–147.

WYBURN, G. M., R. W. PICKFORD and R. J. HIRST. (1964).
Human Senses and Perception. Oliver and Boyd, London.
University of Toronto (1964).

YERKES, R. M. and A. W. YERKES, *The Great Apes* (Yale
University Press, 1929).

YOUNG, P. and E. A. GOLDMAN, *The Wolves of North America*
(Constable, 1944).

ZEUNER, F. E. (1963). *A History of Domesticated Animals.*
Hutchinson, London. Harper & Row, New York (1964).

ZUCKERMAN, S., *The Social Life of Monkeys and Apes* (Kegan
Paul, 1932).